SWIFT AND SURE

First Published in 2024 by Echo Books

Echo Books is an imprint of Superscript Publishing Pty Ltd
ABN 76 644 812 395

Registered Office: PO Box 669, Woodend, Victoria, 3442

www.echobooks.com.au

National Library of Australia Cataloguing-in-Publication entry.

Creator: Peter Evans, author.

Title: Swift and Sure : a career in the Royal Australian Corps of Signals

ISBN: 978-1-922603-66-1 (paperback)

NATIONAL
LIBRARY
OF AUSTRALIA

A catalogue record for this
book is available from the
National Library of Australia

Design by Andrew Davies

echo))
BOOKS

SWIFT AND SURE

A Career in the
Royal Australian Corps of Signals

Peter Evans

CONTENTS

PART I

HALF A MILE IN THIRTY YEARS

From Duntroon to Russell

PROLOGUE

Here I sit in my small and rather shabby office in Defence Headquarters with a commanding view of the rear car-park and the Russell Offices' boiler room. I have a secretary to protect me from unwanted callers, a competent staff to do my work and a 'secret' blue telephone which I salute whenever it rings. I have a wall-full of plaques and testamurs, a grandiose five word title, more initials after my name than in it and wonder at how it all happened.

Thirty years ago I left my parents to be a soldier (or at least a Staff Cadet) and here I am with only a few years left before being compulsorily retired. I've never seen a shot fired in anger and the most warlike thing I've done was to use a borrowed sword to slice open a wedding cake. Nevertheless, I am now a relatively senior officer with a uniform decorated with red tabs and three medals (two for long service) with a sense of amazement at how quickly the years have passed.

I wonder about the seemingly insignificant events that have had a profound affect on my life and those close to me. Conversely, I ponder on the events which seemed so important at the time and which now I barely remember. I think also about my RMC classmates and how few have actually fulfilled the predictions, whether good or bad, made on their futures.

The sense of wonder at where thirty years have gone would not of itself have led me to record my story. I now find that at gatherings of a military kind, especially where members or ex-members of the Royal Australian Corps of Signals are involved, much of the conversation revolves around the 'characters' we have known. The stories told are becoming a type of folklore which is constantly being embellished or distorted in one way or another and, if not recorded soon, will soon be so far away from the truth as to be of little value to those who will follow.

And so I will attempt to set down my own story with reference to as many 'characters' as I can remember. The characters to be portrayed are real and I hope that I can portray them in an accurate way without malice of any kind. If anyone is offended, I apologise, but advise you not to sue unless I happen to sell more copies than the ten required by my immediate family.

EARLY DAYS

Among my earliest recollections are those involving my father's stepfather. He was a most interesting character, having deserted the Royal Navy as a sixpence a week boy – sailor by jumping ship in Sydney. His name was Steptoe but lived the rest of his life under the name of Dickerson and marrying my grandmother under that alias.

I clearly remember, at about age three, sitting on pop's knee while he smoked a pipe and taught me 'dirty ditties'. To increase my enjoyment of such occasions he made me a small toy pipe and so started a habit I retained since being first allowed to smoke at age 17 until I gave up the habit at age 50. No doubt, pop's 'dirty ditties' also laid the foundation of a life-long love of rugby songs!

At the tender age of 4½, I started school with the Good Samaritan Sisters at St James, Forest Lodge and in due course moved next door to the Patrician Brothers to continue my primary schooling. I remember very little of the nuns but not so of the brothers. All but one or two were Irish, stern disciplinarians (one used a golf-bag to carry eight canes which he called Snow White and the Seven Dwarfs) but with a fine sense of humour and a dedication to teaching that was clear to us students, young as we were.

While in primary school, I became an altar-boy having successfully learnt the Latin responses, albeit with an Irish accent. These positions were much sought after as the duties involved the occasional wedding with a five or sometimes ten bob tip or, better still, a funeral which meant riding in a big black car with the officiating priest – a rare treat indeed in the 40s, for a boy from Glebe.

While living at Glebe, I was taken to Holsworthy to see my cousin who was in the RAN and serving his second lot of 90 days for striking a Petty Officer. To the best of my knowledge my cousin and father's stepfather were the only members of the family to have seen military service.

During this time it was discovered that I had a reasonably good voice and I received some basic instruction in singing by the brothers and also at the local Police Boys Club. This never really led me anywhere but instilled a lifelong love of music which has always been a great comfort and given me much joy.

My schooling progressed and eventually I moved up to the secondary school. This was intended to last for three years with an emphasis on English, Book-keeping and Business Principles, so providing a good grounding to becoming a Clerk – a good prospect indeed for a Glebe boy.

At the end of First Year, one of those chance incidents occurred. My father was offered a Housing Commission House in Villawood, a working class outer suburb. This he refused and fortuitously was given a last option to take a house at North Ryde. This he reluctantly accepted as Ryde was at the end of Sydney's longest tram ride and he worked in the city.

Having moved to Ryde, we then found that the Patrician Brothers had another school nearby, Holy Cross College. Having completed First Year I was enrolled in Second Year. After a week or so my parents were advised that I should repeat First Year and take a range of academic subjects which could lead to matriculation. The advice was accepted and I went back a year and joined Billy Stenning who had been a year behind me in Glebe. Probably of more importance was the fact that Holy Cross had a school cadet unit.

Membership in the cadet unit was compulsory but I was excused in my first year due to the fact that there were no uniforms small enough to fit me. Although not a member, I used to stay behind on Fridays to watch the parade and the boys working on their .303 rifles, Brenguns and 6-pounder anti-tank gun. This all seemed so interesting that I began badgering the Officer Commanding, Captain (Brother) John Gallagher to let me join.

After almost a year of lobbying, John relented on the understanding that my mother make good the alterations she no doubt would have to do to my uniforms. After being duly inducted in February 1951, it was found that I was too short to handle a .303 and so was placed in the unit bugle and drum band.

It was soon discovered that I had absolutely no talent for the bugle and so, notwithstanding the probability of tripping on the drum ropes, I began training as a side-drummer. Fortunately, I displayed a degree of natural talent for the drums and my military career was launched.

Under the patient guidance of Sergeant Jack Henry my drumming skills improved and so did my military progression through Lance Corporal to Sergeant in my penultimate year of school.

In my first four years I had become very interested in Chemistry and had thought of a BSc in industrial chemistry if I could get a Commonwealth Scholarship. This aim very much exceeded my parent's expectations and so I received a deal of encouragement to follow this line of endeavour.

During these years, my interest increased in drumming and I joined the Sydney Irish Pipe Band which, incidentally meant joining the Irish National Association, the INA. But more of this later.

At the end of the Fourth Year, I was sent off to an Under-Officer Qualifying Course, where I met Barrie Hungerford and David Gilroy who were to become classmates at Duntroon. I was now vaguely interested in a military career and discussed the possibility of getting into Portsea with one of the Regular Army Instructors. He suggested that, because of my age and the possibility that I would get my Leaving Certificate, I should try for Duntroon and gave me a Prospectus.

I duly qualified at the Course and entered my final year of High School as an Under-Officer with a rather odd single trapezoidal badge on each epaulette and a peaked cap with the edges stuffed with cotton-wool, which was the fashion at the time.

Early in the year we had our annual camp where again I met Gilroy and Hungerford. We shared an interest in Duntroon and after much discussion of our prospects on acceptance (Gilroy was a certainty due to his sporting achievements), the die was cast and I went home from camp to tell my parents of my desire to join the Army. This announcement received a very mixed reception indeed. Mother cried, father was uncertain, my Latin teacher, a pacifist, was horrified but Brother John was excited at the prospect of a Holy Cross pupil getting into Duntroon.

My written application was accepted, I passed my medical and other tests and now faced the final Selection Board chaired by the Commandant, Major General I. R. Campbell. All was going well until lunch when the General asked me what associations I belonged to. When I replied the INA he almost fell of his chair and I had to reassure him that there was no connection with the IRA and, in any case, I only joined to play in the band.

After what seemed an eternity, I received my letter of acceptance giving details of reporting place, what to take, etc. My mother cried again, Dad gave me a lecture on the birds and the bees and I started to pack.

THE ROYAL MILITARY COLLEGE, DUNTROON

The day to leave for Canberra had finally come. I'd had my farewell party, another lecture from Dad, protestations of undying love from my girlfriend, a present from mum (a book called *How Life is Handed On* designed for twelve year olds) and now we were waiting for the train to leave. More tears, a train whistle and my journey to Russell Offices had begun.

There was much excitement on the train among us aspiring Staff Cadets sitting in our First Class (First Class mind) seats and desperately trying to act as sophisticated, men of the world. This was a very difficult task as we were all under 18 and had only left school some two or three months before, however, we tried valiantly.

A fair deal of the excitement had worn off during the six-hour trip and was steadily replaced by apprehension. Just what would it be like? How would we be treated? What did Staff Cadet actually mean? Did we really get leave to come home every month?

At last we arrived in Canberra at about 7.00 pm, from memory. With luggage in hand we alighted onto a dimly lit station to be greeted by a rotund man carrying a long stick with brass on it screaming hysterically at us. Most of us were too frightened to move which seemed to increase his fury. We would have moved but none of us could understand what he was saying and it didn't seem right to interrupt his diatribe to enquire.

One of the chaps noticed a green bus parked outside the station and started moving towards it. This seemed to please the man as his shouting went down a key, so we sheepishly followed and 'em-bussed'.

What happened during the rest of that first day is very blurred, except for a vivid memory of the realization that we had taken a gigantic leap downwards from having been 'somebody' at school. We were hurried through a meal, escorted almost at a run to our rooms and told in no uncertain terms that we were to be out-of-bed by 6. 30 am with our beds stripped and standing outside our doors ready for our shower.

Safely inside my room I took stock. There was a large cupboard with shelves and hanging space, an iron bed, a desk with bookshelves, a chair and an odd-looking fitting on the wall which I soon learnt to be a rifle rack.

On the desk were some new polishing cloths, boot polish, Brasso and some pipe cleaners. Ah! Thought I. Someone had taken the trouble to learn I smoke a pipe – How kind? How mistaken! I soon learnt that pipe cleaners were an essential piece of equipment for cleaning rifle sights. I also learnt that these gifts were not gifts but had been debited to my Cadet Account!

Before I retired for the night, I was visited by a cadet of one year's standing (he having passed into Third Class) who announced that he was my Lord and Master and that he would be responsible for showing me the basics and described in great detail the penalties that would follow if I erred in the slightest way.

Our first day was full of activity. We were attested as Staff Cadets, issued with uniforms and equipment, assigned our College numbers (more of this later) and briefed by a young, slim officer wearing the three pips of a captain and the ribbon of the Military Cross. He was the ADJUTANT to be feared as much as the RSM, the unintelligible man who had met us the night before.

We learnt that our numbers had already been reduced by one. His name was Porter and it seemed that after one meal and a small taste of bastardization he believed that returning home immediately would be his best course of action. I suspect that very few of our Class actually met him but many years later he joined us for a Class Reunion. Bastardization! Here is a word that conjures up many pictures. It has appeared many times in the Press and usually results in investigations, denials, explanations, and interviews with ex-cadets, etc ad nauseam. My memories of its practice are somewhat dim but I recall the theory of it was to bind the Fourth Class together as quickly as possible by making life somewhat unpleasant and also to attempt to identify the senior class 'enthusiasts' by observing their approach to the 'system'.

In broad terms, bastardization seemed to achieve its aims pretty successfully. I never saw any evidence of physical violence and the mental sort was not always minor. We did learn the names of the Corps of Staff Cadets very quickly, as we did the inscription on General Bridges' grave and all manner of essential facts, such as the height of the flagpole. We also learnt to quickly come to grips with the reality of our situation as being on the very bottom rung of the military ladder.

FOURTH CLASS

We were indeed Fourth class, for the first few weeks we were referred to only as Fourth Class. We had no names we were just Fourth Class.

After the settling in period we were all interviewed by the Director of Civil Studies, Professor Traill (The Bum) Sutherland. This was a most important event as he determined what course of civil studies we were to undertake. It seemed that the major criterion was knowledge of mathematics. This was quite understandable as the good professor's subject was mathematics.

There were three options available, Engineering, Science or Arts. These were in descending order of preference (The Bum's) depending on the level of mathematics studied at school. As a result there were many unwilling entrants into the intellectually demanding areas of Engineering and Science. Some of the unwilling managed a swift movement through Science to Arts, often with the assistance of the their Lords and Masters.

I had gone into the RMC with a vague intention of studying engineering, so I was very pleased with having been allocated to the C class. I have to admit that I enjoyed studying engineering and so was mildly successful in the academic side of things.

After a month or so, and after our 'initiation', we were advised that it was time for us to be launched onto the unsuspecting Canberra social scene. The vehicle for this was the Fourth Class Tennis Party. As none of us had been outside the gates since our arrival (being considered as unfit to be seen in public wearing the uniform of the Corps of Staff Cadets), there was the minor difficulty of finding a partner for the event. This problem was largely overcome for some of us by our Lords and Masters matching us with suitable partners – for the rest it was pot luck.

The party consisted of games of tennis followed by a dance in the evening where we were all heavily chaperoned. Music for the dance was provided by the College jazz band. This was a traditional Dixieland jazz band known as the Civic City Seven. I was enraptured by the music and, I'm afraid, gave scant attention to my partner. This was just as well as my partner was the girlfriend of my Lord and Master and he had threatened me with dire consequences of the mildest flirtation on my part.

At the first opportunity, I made myself known to the band leader, Lachie Thompson, and sought to join as the drummer. As luck would have it, the incumbent, David Wise, was keen to follow other pursuits and I soon found myself part of the band. The transition from pipe band to jazz band was made easy by Lachie and I became quite proficient in the rhythm section.

The band was a major preoccupation during Fourth Class leading to a comment on the end of year report to my parents along the lines:

'If Staff Cadet Evans were to spend on his studies half the time he spends on the jazz band, he would probably do well!'

The band was both my salvation and great temptation!

THIRD CLASS

On returning to the College in 1956 we had the pleasure (but not immediately) of being elevated from the lowest form of life to being the higher of the lower. This promotion was not evident, of course, until the new Fourth Class arrived. Life became easier although this was really just a matter of degree. We still faced the wrath of a variety of people inside and outside the Company. These included our Section Corporals, our Platoon Sergeant, the Company Sergeant Major, our Company Commander, Drill Sergeants, the Regimental Sergeant Major, the Adjutant, Commanding Officer of the Corps, Physical Training Instructors and a variety of military skills training staff. To this formidable list must be added (for those studying mathematics) 'The Bum'. I stand to be corrected, but I believe the 'The Bum' handed out more extra drills (for late homework, etc) than most of the military staff. I also believe we received from him a great grounding in pure mathematics that was to serve many of us well.

The band changed dramatically with the graduating class of 1955 removing all the 'frontline'. Brian Oxley continued on piano with myself on drums as the mainstays. An attempt was made by a member of First Class to introduce an electric guitar but Brian and I managed to derail this. The music was very different but I still found it a great source of enjoyment.

At the start of winter, I took up a new sport, Hockey, having made the decision that Rugby was best left to the taller members of the Corps. I seemed to have a fairly good flair for the game and became a competent forward. I played in the Thirds all year but moved to the Firsts in 1957.

Life was proceeding pretty well when we experienced the disaster of the drowning on Lake George in July. Much has been written about this event in other books and I will not attempt to expand on these accounts. What I can say is that the death of five of our classmates had a profound affect on the Class and that is still evident to this day. We who were to become the Class of '58 became an extremely close-knit group that developed a substantial capacity to look out for each other whatever the adversity.

As the reader would imagine the Lake George drownings were of such significance that I can remember little of events later in the year.

SECOND CLASS

Second Class now meant that we had progressed to the lowest of the high – a great improvement in status and pay – now 18 shillings per week!

My studies were going well and I had advanced to the First Hockey eleven playing either on the wing or centre-forward. Late in the year the Firsts travelled to Sydney to play in a match against Sydney University. On this trip I met Yvonne Cotton and a whirlwind romance followed although we only met on three or so occasions before I went on leave for Christmas.

Second Class was of great importance, as we would need to decide on our future Corps for our military service after commissioning. Notwithstanding my desire to complete a degree in electrical engineering, I was attracted to the idea of serving in the Royal Regiment of Australian Artillery as were David Gilroy and Barrie Hungerford and others. David, Barrie and I had met during school cadet camps and they were my oldest friends at the College.

My interest in becoming a Gunner led to my volunteering to be in firing parties for gun salutes which were at that time fired from Mt Pleasant behind the Australian War Memorial. I have a reasonably clear memory of being in one of these saluting parties, under the command of Captain Clive Simpson, Royal Horse Artillery, for the opening of Parliament by the Governor-General Sir William Slim.

I imagine the arrangements for saluting parties have not changed to this day. Four guns are lined up with the appropriate store of ammunition next to each gun. The gun next and second next to be fired are loaded so that, in the event of a misfire, the second gun can be fired with only a minimum of delay. If a misfire occurs, ammunition at the guns is rearranged and an extra round is extracted from the reserve to cover for the misfire.

On the day in question all was organised and all was well. The salute started on time and the interval between shots was spot-on. Then the misfire! The drills seemed to work well and firing continued. Unfortunately, in the enthusiasm of the moment to ensure that the misfire was covered TWO extra rounds were delivered to the guns. This resulted in a Twenty Two gun salute, stern words from the G-G, duty officer for the unfortunate Clive and extra drills for we prospective Gunners – not a good omen.

Towards the end of the year the whole of Second Class plus the members of First Class going to the RAA went off on the annual Artillery Trek. At this stage I was still keen to be a Gunner, being unfazed by the saluting fiasco, and was part of the battery command post staff.

At night in a gun position, the guns are laid (aimed) on the most likely target so that gunfire can be bought to bear with minimum delay after receipt of the request for fire support. One gun crew member stays with each gun ready to fire if the appropriate code word is shouted from the command post. As soon as a gun is fired, the gun crews race to their positions and all guns engage. In theory, a very simple operation which has been effectively carried out over many years.

Early in the exercise, I was taking part in an unauthorized party in the command post in the early hours of the morning. I left the tent to answer a call of nature, tripped over a guy-wire and uttered an expletive louder than I had intended. A class-mate (who actually became a Gunner) on the gun mistook my expletive for the code word and fired. In best Gunner tradition the remaining guns of the battery were fired. This event caused some consternation among the Artillery instructors and I was subjected to some sharp criticism. Coupled with the previous problem of the 22-gun salute, I thought I should review my Corps ambitions.

When the time for Corps selection came, I had pretty well made up my mind to go for the Royal Australian Engineers as I quite enjoyed the mathematics involved in structural design and I had mastered the thrilling 'slump test' for concrete. On leaving the room having given RAE as my first choice, I was stopped by the Signals Instructor, Major Reg Williams – an absolutely charming man, who asked me about my choice. When I said RAE he replied that I should really consider the Royal Australian Corps of Signals as they would love to have me. The invitation was too much for me to resist so I promptly 'About Turned"'re-entered the room and elected the Blue Lanyards as my first choice. This decision, made on the flimsiest of reasons, I never regretted.

FIRST CLASS

Back at 'Clink' with less than a year to go before graduation. We were now the highest of the high in cadet terms while having no real standing in the Army as such.

I was promoted to Corporal. I was not given a Section to command but instead was made President of the Quarter Bar Committee – an appointment I still hold and which enables me to preside at Class reunions.

Until First Class, we were not allowed to drink in the ACT unless in the company of parents or guardians. In 1957 the powers that be reconsidered this policy and the Quarter Bar was established next to the Officers Mess in Duntroon House where we could regale ourselves on beer for an hour or so on three or four nights a week. This was an enlightened view as at the end of the year we would be catapulted in to Officers Messes with little or no experience of handling the demon grog.

I'm afraid I did not handle the authority given to First Class over the junior cadets at all well. As is often the case the bullied become the bully and I made life miserable for a few. In the unlikely event that any of my victims should read this book, I offer my heartfelt apologies.

First Class was a time when cadets were given some specific Corps training to fit them for their first postings. For those in the Engineering Class, our specialist training was mainly in academic work in the particular engineering stream we were to enter. We who were destined for RASigs, Bob O'Neill, who became a Rhodes Scholar, Steve Hart and myself, did manage to spend some time in the Signals Hut with Major Williams and Warrant Officers Harry Hutton and Tom Fitzpatrick.

In the hut we were given a grounding in radio procedure and the tactical uses of radio using WWII vintage equipment. We also learnt something of the niceties of cable laying, switchboards, antenna types and construction. More importantly, we were imbued with a sense of *esprit de corps* and a love of the Corps of Signals, a love that stayed with me throughout my service.

The Cotton/Evans romance continued to develop to the stage where I borrowed £90 from my Grandmother and bought an engagement ring. Arrangements were made for an engagement party to be held at Yvonne's house during the mid-year leave. Just before going on this leave the Class was subjected to a lecture on the dangers of an early marriage and a recommendation that we should observe the old Indian Army maxim:

"Lieutenants will not marry, Captains should not marry, Majors may marry, Colonels will marry."

Unfortunately, the engagement party received some minor press coverage and so came to the notice of the RMC authorities which led to me receiving some unwanted policy advice and some reasonably severe bollocking!

The rest of the year flew by and soon it was the second Tuesday in December – Graduation Parade in the morning and Graduation Ball that night.

Wednesday morning came and we were off to the Sergeants Mess as the newest lot of Lieutenants in the Australian and New Zealand armies. What a sight to behold! All of us a bit bleary eyed and all suffering from neck cramp as we admired the bright shiny pips on our shoulders. We were on the verge of our careers, full of the joy of life and ready to do what was necessary to find the elusive Field Marshal's Baton that was allegedly hidden in every defaulter's pack.

After leave most of the Class would be off to Young Officers' Courses at their respective Corps Schools after which many would go to National Service Battalions. For those of us in the 'C' Class that had succeeded in passing our finals a different path beckoned – off to academic institutions to complete degrees or diplomas in engineering.

So at the end of 1958 I had graduated 10th in my Class from RMC and had been commissioned as a Lieutenant in the Australian Staff Corps and allocated to the Royal Australian Corps of Signals. At the stroke of midnight on 9 December 1958 I had become an officer and a gentleman with an annual salary of almost £1000. I was twenty-one years old but in reality had not reached a level of maturity that might have been expected of someone who had already been away from home for four years.

UNIVERSITY OF NEW SOUTH WALES

As I had maintained the necessary average marks during my four years in the 'C Class, I had sufficient credits to enter directly into Third Year of a four-year degree course. I had hoped to go the Sydney University but a senior academic at that institution had claimed "*the transistor was a fad and that Sydney University would continue its emphasis on valve technology*". This did not please the then Director of Signals so I was sent to what was then called the University of Technology but which was renamed the University of New South Wales before I graduated.

At the completion of Christmas leave I reported to the office of the Chief Signals Officer, Eastern Command, where I was to be attached until I had completed civil schooling as it was called. There were many formalities to be completed. I obtained my Pay Book, received briefings on my responsibilities as a student, was advised of the book and stationery allowance (an extremely generous one I must say) I would receive, etc. I then presented myself to the University to enrol.

Initially the enrolling process was quite simple. I presented documentation from RMC, was duly granted exemptions for the first two years of the course and I stated my intention to join the communications stream of Electrical Engineering. I then hit my first snag – the University required that I become a member of the University Union but the Army was, as one would expect, totally opposed to unionism of any sort. The staff officer responsible for training in general told me I could not join whereas the University was adamant that student union membership was a necessary requirement for attendance. There followed some intense negotiation that ended when I undertook to pay my own fees (the only cost to me for my tertiary education) and to refrain from taking part in student union demonstrations. The whole thing was really quite ridiculous as there had been many before who had moved from RMC to tertiary institutions without difficulty – I expect I may have run up against a disgruntled staff officer who had an antipathy to RMC graduates.

Having now got through the administration, I thought it was time to present myself to the Chief Signals Officer (CSO), Lieutenant Colonel I. J. Hooker. CSOs were Lieutenant Colonels in the major commands and Majors in the lesser ones. They wielded a great deal of power within their commands and commanding officers, even though of the same rank, tended to demur to them. At the appointed time I was ushered into the great man's office not really knowing what to expect. I certainly did not expect his opening remarks.

Without preamble, he asked:

"Do you have a Mess Kit (the Army equivalent of white tie)?

"No Sir" says I.

"You may use my tailor – H G Newton and Son. Come back and see me after your first term. March out."

Mess Kits were not much seen in the late 50s as most officers were content with the Blues uniform. However, I thought discretion being the better part of valour, I should avail myself of the services of the said H G Newton. I called on the firm and was informed that HG himself would not be able to cut my uniform as he was over-extended making kilts. His son would be tasked with the job, but as he only had 25 years experience, HG himself would supervise. Measurements were taken and after a number of fittings my Mess Kit, complete with overalls, was ready for collection on payment of the princely sum of 100 Guineas (more than one month's pay) just in time for the Annual Signals Dinner.

In 1959, the campus was still at Ultimo near the Technological Museum. From memory there were only Science and Engineering faculties. The real expansion of the university was to occur after the move to Kensington. The teaching facilities were pretty tired but we did have excellent teachers and reasonably modern test equipment.

I lived with my parents at Ryde and drove daily into town. There was no student parking so it was necessary to leave home early to beat the traffic and secure a park in the surrounding streets.

I joined my new class as the new boy knowing only Peter Wilkins who had graduated a year before me and was now starting his final year. I was then surprised and pleased to find that two of my new classmates had been at school with my fiancée. It was chalk and cheese from Duntroon although I soon found that we were expected to wear 'uniform' to class. This consisted of shirt and tie, desert boots, corduroy trousers and a tweed sports coat.

Lectures went from 9 to 1 each day, practical work from 2 to 5 with Wednesdays off for sport. There was little choice in core subjects with emphasis being placed on pure and applied mathematics, physics, circuit theory, etc. Philosophy was compulsory and one other humanity subject had to be taken. I elected to do Sociology and I carried this on in Fourth Year. The two humanities were foreign to me but I found that I liked the subjects and did well in them.

I soon found that the curriculum left little time for involvement in student activities and I limited myself to hockey during the winter months. I was anxious to do well academically and spent most weeknights working till midnight on lecture material and writing up practical assignments.

I did have a tremendous advantage over most of the other students as I was being paid my salary as a Lieutenant plus a generous book allowance and access to the Army stationery store. In contrast many of the students were on cadetships or scholarships and had to scrimp. From memory there was only one member of the class who was a fully private student funded entirely by his parents. I made friends with a number of my classmates but these friendships were not on the same level as those I had made at RMC.

I very much enjoyed the challenge of mastering new subjects and being exposed to new technologies. The transistor was just eleven years old when I started Third Year so we learnt about these new-fangled devices by analogy with thermionic valves. I did well in Third Year but was by no means the top student. I was required to report to the CSO from time to time to report on my progress but generally speaking I had very little contact with the Army during the year apart from the Corps Dinner mentioned previously.

The Corps Dinner was held at the old Imperial Service Club in Barrack Street, Sydney. The Club was a marvellous place with lots of oak panelling, silver, portraits of long dead members and club servants who appeared to be way past the age when one should be working. The Club was Rum Corps through and through and very much the province of the citizen military. I found most of the members polite but I had the distinct impression that the regulars were barely tolerated and there was certainly no great effort made to attract regulars to Club membership.

I do have a clear memory of asking one of the ancient club stewards what time the bar shut. He replied in a tremulous voice:

"The last time, Sir, was 1914 when the war started."

The practice of running a twenty-four hour a day bar was to continue even after the Club's disastrous move to York Street and, no doubt, affected the Club's viability.

Back at the University, the challenges continued and I found that I needed to spend five or more hours a night studying and writing up practical work. I did leave the weekends free as far as I was able to spend with Yvonne. We had been engaged for over a year but the original intention was that we would not marry until I had finished my degree. The waiting became more and more unpalatable so we decided to bring the wedding forward to January 1960. As was the norm in those days, I wrote to my Commanding Officer, the CSO, seeking permission to marry. This was grudgingly given but I was left in no doubt that I would not receive any assistance with housing. The reaction from my future mother-in-law was quite unexpected. She now bitterly opposed the marriage and as Yvonne was 19, I would need permission from the Courts to go ahead. After much cajoling interspersed with threats of the Court permission was given and planning for the wedding started in earnest.

I had maintained a close contact with the Brothers at my old school and the Principal, Brother John Gallagher, had determined that our marriage would be the first to be celebrated in the newly completed College Chapel. I had to apply a modicum of braking on the enthusiasm for the wedding on the part of the Brothers as Brother John was toying with a Guard of Honour being provided by the whole school! We settled for a Nuptial Mass for the morning of 9 January 1960 with two of the Brothers acting as servers and with a guard provided by eight of my Sydney-based classmates.

Part of the requirement for an engineering degree was completion of an appropriate period of time working as an engineer in some public or private enterprise. Here the CSO was most helpful as he arranged for me to spend the necessary time at the AWA factory in Strathfield. I was to be given a range of experience in the design laboratory, quality control and manufacturing. Apart from completing the degree requirement, I was also able to make contacts with the management and engineering staff that would be useful later in my career.

I received my first experience in electronic design in the laboratory. I was given the task of designing a small transistor amplifier for a mine telephone. I immediately got out my notes, a few reference books, my slide rule and got started. After some hours I had not got too far when one of the old hands said that I had the approach all wrong. He advised connecting variable resistors across the transistor terminals, adjusting them until a gain was obtained and then measuring the values of the resistors. Hardly scientific but a useful practical approach although one that would receive little favour among the faculty staff!

My next attachment was to quality control in the commercial (radios, radiograms and TVs) electronics section. I made my first discovery about commercial matters when I saw that several different radiogram models were being produced with identical electronics and similar cabinets but with different model names and vastly different prices. When I sought an explanation for this I was told that there were many customers who liked to tell their friends how much they had paid for their purchase and that AWA was simply reacting to the demands of the market. I guess nothing much has changed.

My final placement was in quality control where I was required to wear a grey dust coat. One of the workers asked the foreman, a bit of a wag, who I was. He replied that I was a foreman in training. Given my extreme youth this caused a major problem and a strike was only averted by the foreman coming clean about my attachment from the University.

And so ended 1959 – I was well on the way to acquiring a degree and a wife but I think I was still not as mature as I should have been.

1960

9 January was quickly upon us and the Holy Cross College Memorial Chapel had its first wedding. This was in the days in the Catholic Church that required one to fast from midnight before receiving Holy Communion. For reasons that I cannot remember, we wanted the reception in the afternoon and so the Nuptial Mass was timed to commence just before mid-day, the latest time that a Mass could start.

My best man and groomsman, Rod Stewart and John Wertheimer, were both non-Catholics and were therefore not used to genuflecting in the normal course of events, let alone when carrying a sword. This lack of experience and rehearsal was an early source of amusement to the congregation as one of my party managed to get a sword between his legs while genuflecting and only narrowly avoided spread-eagling himself at the altar rails.

At the conclusion of the Nuptial Mass, the wedding party retired to the Sacristy for the signing of the Marriage Register. The Brothers had very thoughtfully provided tea and toast for the bride, bridesmaids and my attendants. They considered it more appropriate for me to have something stronger, so a whiskey (Irish of course) had been discreetly placed where I could steady my nerves. Brother John insisted my nerves should be really steady, so a second nip was needed which on an empty stomach had me floating on air!

Off we all went to a reception centre on the North Shore and from there Yvonne and I drove to a hotel in Parramatta en route to our honeymoon at the Hydro Majestic in the Blue Mountains. The 'Hydro' was quite an institution in the Sydney of the late 50s being well known for the number of couples named Smith that stayed over weekends.

With the honeymoon over, it was back to Sydney where I was attached to 402 Signal Regiment in Dundas for a few weeks until University resumed. 402 Signal Regiment was a major station in what was known as AUSTCAN (Australian Communications Army Network) and which provided access into the allied communications networks. The Regiment consisted of a message-switching centre at

Dundas and Transmitting and Receiving Stations at Wallgrove and Bringelly respectively.

The Commanding Officer was one Lieutenant Colonel DAC Griffith who had transferred from the British Army and was known as Handkerchief Charlie due to his practice of carrying a handkerchief up the sleeve of his jacket.

This was my first experience of a working unit. I was intrigued to find that everything seemed to work well without too much fuss or bother. My duties were minor but I set about learning what I could of the technical operations of the Regiment. My endeavours were interrupted on a regular basis as word would come down from the Adjutant that the CO had guests coming to the Mess for lunch and all officers were expected to appear in their Service Dress uniforms. This often meant a quick trip to the house in Gladesville where Yvonne and I rented a room and back to Dundas. It was a very pleasant existence indeed and something I could not have imagined while a cadet at Duntroon.

We dined formally about once a fortnight in our tiny Mess. We would gather in the anteroom for sherry and then proceed into dinner. The dinner itself was a subdued affair and all were required to remain at the table until after the Loyal Toast. This could be quite a challenge in bladder control! After dinner we often played mess games under the watchful eye of the PMC and the CO.

The CO was fond of challenging officers to perform a balancing act at which he was expert. Three legs of a chair were balanced on beer bottles and the competitor was required to approach the chair from the back and place one hand on the seat and one on the back. One then had to balance, reposition to sit on the chair, lean forward and recover a glass of beer from the floor and drink it. For the life of me I cannot remember how this was done but I did become quite good at the trick.

The CO was fond of champagne and often called for it to finish off a dining-in night. At one of the first of the dinners I attended, I approached the CO to see if he required a top up. I made the grave mistake of asking him if he wanted 'another' drink. I was advised that 'a gentleman's drink is always his first' and I should have asked 'if he cared for a drink'. The lesson in etiquette was reinforced when he made me Duty Officer for a week. This lesson has stayed with me to this day.

Soon this idyllic life was over and it was back to the University. As I had done very well in my Third Year examinations, I was invited to enrol in the Honours programme. This meant returning to the campus two nights a week to take additional subjects so there was even less time to take part in student activities. A humanities subject was required and I elected to continue with Sociology.

By this time the University had been renamed the University of New South Wales and the move to the Kensington campus had started. This meant that we had the inconvenience of having lectures split between the old campus at Ultimo and the fledgling campus at Kensington. The building programme was well underway but we still had some classes in temporary wooden huts. The move also generated a new nickname for the University – Kenso Tech. We were very much 'red brick' compared with Sydney University but we had the advantage of a progressive faculty that introduced us well to the emerging semiconductor technologies.

Yvonne had become pregnant early in the year and we had found it necessary to find better accommodation. We tried sharing a house in Ryde and then moved to Eastwood where we had half a house to ourselves. Yvonne had found a doctor very close to the Ryde Hospital where she was booked for the delivery.

On the morning of the 17th October, Yvonne was feeling unwell so I excused myself from class and presented Yvonne at her doctor's surgery. She had barely got inside when 'the water broke' and I was told to get Yvonne to the hospital ASAP. We had been in the hospital only a few minutes when Yvonne produced Damian. It may not have been the shortest labour in history but it certainly seemed so to me.

As the end of the year approached, I began to wonder about the next step in my career. I had done very well in Sociology and my lecturer had suggested I undertake post-graduate work in that discipline. I mentioned this to the CSO who went apoplectic and poured gallons of cold water on the proposal. I was still interested in doing post-graduate work so I was sent to Melbourne to see the Army Chief Scientist. He was sympathetic to the proposal but said he would only recommend it if I obtained First Class Honours. This was unlikely and I must say I was somewhat relieved as I thought it time that I did some soldiering.

Even though examination results would not be available until the New Year, I had been told that I would certainly get my degree. After advising the CSO of this, I received a Posting Order which appointed me as the Troop Commander of 2 Radio Installation Troop of 1 Signal Project Squadron based in Watsonia Barracks, Melbourne.

And so ended 1960.

DIGGERS REST

Early in 1961, I headed to Melbourne to report to my new unit and start my first appointment.

1 Signal Project Squadron was located within Watsonia Barracks with limited office and drawing office space and a rudimentary workshop located in 403 Signal Regiment's vehicle lines. The unit was commanded by Major Ian Tee who was a WWII veteran, as were most of the officers of the Corps. The preponderance of members of the Squadron were technicians and draughtsman with many being at the top end of their trades. There was not much spit and polish in the Squadron but it did enjoy an enviable reputation for doing installation work of the highest standard. The Squadron also had a reputation for producing 'Go-Karts' and repairing the lift equipment for a popular ski run.

There were many projects in the pipeline at that time. Among them were new transmitting stations at Diggers Rest and Wallgrove and receiving stations at Rockbank and Bringelly. From memory there were also a number of major antenna construction projects to support the Army's fixed communication network.

After a short period of orientation at Watsonia, I travelled out to Diggers Rest to take command of 2 Radio Installation Troop. We were tasked with completing the installation of the 'new' transmitting station. The 'old' station on the same site consisted of WWII vintage transmitters connected to the rhombic and dipole antennae in the antennae farm by a very dangerous but flexible manual switching arrangement. SWB 8 and SWB 11 transmitters were to be replaced by the AWA built E513 and E514 10 kW transmitters and the Marconi 30 kW E10. More importantly, in my view, was the installation of the Antennae Switching System that removed the possibility of an operator coming into contact with high-energy radio frequency signals. Work had been in progress for some time but was hampered by lack of cabling and a shortage of manpower.

My 'command' consisted of a WO2 Foreman of Signals, a Staff Sergeant Foreman of Signals, a number of other technicians, a Staff Sergeant looking after stores and a Driver/Batman – a total of ten, I think. The WO2 (much later to become Lieutenant Colonel) was Jim Messini who became a firm friend and was a great help in my learning how to apply theoretical training to practical engineering. I am also indebted to WO 'Mac' McCoubrey who taught me much on how to become an effective officer. I have always felt that one of the great distinguishing features of the Australian Army is the great relationship that exists between Warrant and Commissioned Officers and these two were particularly good examples of men who were prepared to encourage, correct and mentor while still recognizing the seniority in position of much younger men.

Jim ran the troop with my main responsibilities being to shelter the Troop from the Squadron, to liaise with the station commander of the 'old' station, to conduct visitors around the installation and to write the monthly Progress Report. I had hoped to immerse myself in some of the technical issues of the installation and so, from time to time, put on overalls and crawled around the place. This was stopped when the OC made a surprise visit and caught me at it. I was told in no uncertain terms that I should on no account be dressed in working dress and that I should concentrate on the 'higher functions'.

My troop was an interesting lot and my RMC education had not done much to prepare me to deal with eccentrics. My Staff Sergeant Q was quite paranoid about being the target of unnamed conspiracies to 'get him'; one of my Sergeants had an unswerving self-image of being a modern-day Casanova and always kept a condom in his wallet in case of emergencies – unfortunately he also had a taste for strong drink which generally prevented him fulfilling other ambitions; and my Driver/ Batman spent all of his spare duty and leisure time in polishing every accessible surface of the Troop utility. Notwithstanding the preceding, they all worked like Trojans and real progress was being made in bringing the station to completion.

Early in the year I managed to find rental accommodation in Airport West and I moved Yvonne and Damian down to join me. Married Quarters were less than plentiful and I did not have many points to claim a priority.

I was still coming to grips with living in Victoria. The early part of the year was alright as it was still summer. Then with the change of season came Australian Rules Football. This game was barely tolerated at RMC and was held in even less regard than Hockey. I did not understand the game and found it hard to cope with the papers and TV being full of it. I was even less enamoured of the game when, on one Saturday afternoon while walking to the local Tobacconist wearing my RMC muffler (white with each end bearing a red and navy blue stripe separated by a thin black line), I was accosted by a group, somewhat the worse for wear, leaving the local pub who threatened to thrash me and any other Essendon supporters in the vicinity. They weren't really interested in my explanation that RMC shared the colours with Essendon and I was obliged to make a rapid and undignified retreat. I never have learned enough about the game to make it comprehensible and should put acquiring such knowledge on my retirement To-Do List.

By mid-year we were able to connect some of the transmitters to dummy loads so that testing could commence. Progress attracted attention from the Directorate of Signals and we were subjected to many visits. During such visits we would fire-up as many of the transmitters as feasible to impress the visitors. I would equip myself with a neon tube to use as a pointer. On pointing the neon tube to a running transmitter, the tube would like up in spectacular fashion due to the presence of high-energy radiation. This was my party trick but I do not claim its invention. I sometimes wonder what effects the constant radiation levels had on the long-term health of members of the Troop.

I mentioned earlier that one of my main tasks was preparation of the monthly progress report. This was always a struggle as I did not have access to the original project plan so how was I to measure progress? The answer was that I really could not report progress but all that I could do was report on the amount of effort expended and problems that may have or were likely to appear. Being a good soldier, I did what I was told and produced well laid out reports with good staff duties. The content was very questionable but I was never questioned which lead me to wonder if the reports were ever read.

I was learning how to be a leader using the tried and true methods instilled at RMC – the three 'Fs': Fair, Firm and Friendly. It was good experience and I was enjoying myself. My enjoyment was marred in a way by the visit of a senior officer from the Directorate who was concerned about the time being taken to bring the station on-line. After his inspection he lined me up and directed that the station be completed by the end of the year irrespective of manpower or materials! I hasten to add that this sort of rather idiotic pronouncement was uncommon.

September marked a new era in the Squadron with the arrival of Major David McMillen to be the new Officer Commanding. David was one of the new breed of officers in the Corps. He was a RMC graduate of the Class of '48 with a BSc degree. He was a charming and cultured gentleman with an engaging personality. We became firm friends and I had the good fortune to serve under him as a Lieutenant, Captain, Major and Lieutenant Colonel. David will appear many times in this narrative.

After a few months I moved the family from Airport West to a much nicer house in Niddrie and made friends with some neighbours. Yvonne was pregnant and the baby was expected in October. We had booked into the maternity wing of the Sacred Heart Hospital on

Moreland Road in Coburg and all preparations had been made. A little earlier than expected, Yvonne announced that her labour had begun. Having had the experience of Damian's birth I took things calmly, located the hospital bag, rang my mother in Sydney to warn her to come, parked Damian with a neighbour and then changed into suitable attire. Some time had elapsed and Yvonne informed me that labour pains were becoming more intense and with shorter gaps between them. We climbed into the car and it was only then that I noticed that a heavy fog had descended and my calmness evaporated. I managed to flood the carburettor but then got going. It was a journey that I will not forget. The fog did not help and things were getting pretty dicey by the time I got Yvonne into the waiting room at the hospital. I left to park my car, returning in a few minutes to be informed that a daughter had been delivered and that mother and child were being moved into a Ward and I would be able to visit shortly. Notwithstanding the above circumstances I was given a bill for Theatre Fees!

My daughter, Kerry Ann, was the first girl born into the Evans family in three generations and was duly doted on, particularly by my father.

SYDNEY

1962 started with a Posting Order from 2 Radio Installation Troop to Second in Command of the Squadron. This appointment was not really in the command chain but was more in the nature of a project co-ordinator. I fully expected that this would be at least a twelve month posting, so I took the step of ordering a new car with a bigger boot to handle the paraphernalia of two small children. I had no sooner placed the order when I received another Posting Order moving me to 402 Signal Regiment in Dundas on promotion to Temporary Captain. The move was inconvenient but welcome as it brought us back to Sydney to the grandparents and put some extra cash in our pockets. The short notice was due to the fact that I was required to do an in-house course on STRAD (Simultaneous Transmit Receive And Distribution) before taking up the appointment.

STRAD was the acronym for an automatic, hard-wired, computer-based telegraph switching system designed and built by Standard Telephones and Cables (UK). Two were purchased for the Army network and were based in Watsonia, Victoria and Dundas, Sydney. This equipment was at the very forefront of technology and amounted to a quantum leap in efficiency when compared to the torn-tape systems generally in use in both the commercial and military communication systems. I remain in awe of the courage shown by the officers of the Directorate of Signals in the late '50s to commit to such a risky procurement.

At the completion of my STRAD Course, I took up my appointment as the Chief Systems Control Officer for the Regiment. This grand sounding title translated to engineering officer and I was responsible for the technical operation of the switching equipment and the radio relay connecting Dundas to the message centre at Victoria Barracks and the Transmitting (Wallgrove) and Receiving (Bringelly) Stations.

The Regiment was commanded by Lieutenant Colonel Roy 'Deafy' Lawrence with Major Stan Blackburn as 2IC. Majors 'Big Jim' Little and Jack Seager were the two Squadron Commanders and Captain Eric Swann was the Quartermaster. All these officers were WWII veterans and all had been commissioned from the ranks. This was very much the norm in the early sixties as there were few OCS or RMC officers in the Corps.

The Regiment's main location on Kissing Point Road, Dundas had been acquired from a commercial broadcasting station. It was an ideal location for a switching centre with its 150-feet tower providing a clear line of sight to Victoria Barracks and the outstations. The move into Dundas had been completed in haste and the available buildings were bursting at the seams. The technical portion of switching centre, workshop and cryptographic office occupied the old broadcast station in what was known as the 'White House'. There was some limited living accommodation, a small building with the Officers' and

Sergeants' Messes at either end with a common kitchen in the centre and an administrative area for regimental headquarters. The latter was overflowing with cupboards. Even a toilet had been converted to office space. These close quarters led to many humorous events, some of which will be related later.

The 'White House' had little spare space, particularly while STRAD and the old manual torn-tape system had to operate in tandem while the new system was being trialled. I shared a longish narrow office with my desk at the front near the door protecting 'Big Jim' whose desk was behind mine. Next along the corridor was the Crypto room that was heavily locked with access strictly controlled to those with suitable security clearances and legitimate reasons to enter. Next was a large space filled with STRAD, torn-tape send and receive machines, the radio relay racks with a few sundry small offices including space for the STC contractors. For security reasons the front door to the 'White House' was equipped with a keypad lock and a very temperamental automatic door-closer.

402 Signal Regiment was the second most important relay station in the AUSTCAN (Australian Communications Army Network) and it also played an important part in the COMCAN (Commonwealth Communications Army Network) in that it provided a major link to the UK via Nairobi. The Regiment also provided a number of gateways for field units on deployment and occasional connections to the other Services.

Communications was our major function but this was now a peace time Army so administration, especially accounting for stores, was a high priority for Commanding Officers. Our CO was no exception and I recall my initial interview when I was told that loss of messages through technical failures would earn me his wrath but losses from my technicians' toolboxes would earn me a court martial!

So I started work in the 'White House' and adapting to my new rank of Captain. I guess most armies are the same in their treatment of young officers; in our case Lieutenants are considered by the troops as apprentices and are well looked after whereas Captains are on their own although guidance, in one form or another, is often forthcoming from the Warrant Officers.

In the first few weeks in the 'White House', I found it necessary to engage with Warrant Officer Joyce Curran over some matter that utterly escapes me now. Joyce took umbrage and delivered the best put-down I have ever experienced. She looked me up and down and then said in tones that I will leave to your imagination:

"Listen Sir, I've had more leave than you have had service. Why don't you rack off and leave me alone."

She was absolutely right of course as she had been around for a long time and had the medals to prove it. Armed with advice I had received from Harry Hutton at RMC on dealing with formidable female Warrant Officers, I replied:

"Good idea"

and retreated to the Mess.

Many, many years later when I was Representative Colonel Commandant and guest of honour at a Signals reunion, I met with Joyce and reminded her of the above. With a twinkle in her eye she denied all knowledge but I thanked her for the good lesson she gave me.

I had had little experience in the operation of Messes in my former unit so the CO appointed me Secretary/ Treasurer of the Mess Committee. Mess books of account were usually maintained as simple single entry systems but I was encouraged by Jack Seager to use a double entry system that had many advantages in finding balance, etc. The major disadvantage was that the CO was learning accountancy for a post-retirement occupation and he insisted on reviewing the books every month and correcting what alleged errors he found. He was usually wrong so I often found myself re-doing accounts on a regular basis. The experience did me no harm as the coaching from Jack Seager was the only financial training I ever received.

My Secretary/Treasurer appointment was to get me into some serious trouble with the CO not long after I arrived. This was due to the fact that the Quartermaster, Eric Swann, who was also a senior wine judge and well connected in the wine industry urged me to make a bulk purchase of vintage port from All Saints. A couple of barrels of 1937 port had been discovered and at Eric's urging I purchased them for the Mess, had them suitably bottled and forwarded to Dundas.

A week or so later, the Adjutant came rushing to find me and escort me, post haste, to the CO. The CO had been examining the Mess books and found that we had temporarily gone into debt. The exchange, I couldn't call it a conversation, between the CO and me went something like:

"Peter, we are in the red."

"Yes, Sir," I replied standing rigidly to attention.

"Why?"

"I bought some Port"

"How much?"

"One hundred cases."

Given that there were only eight or so officers in the Mess his perturbation was understandable so I hastened to assure him that the purchase was a sound investment. I was subjected to a stern lecture on the evils of young officers overstepping their (limited) authority and I was made Duty Officer for a week to learn the lesson. The lesson I learnt was to beware of COs studying accountancy!

I now wish to leave the Regiment temporarily to describe my domestic circumstances. As soon as I had arrived in Sydney I began the search for suitable accommodation. Married Quarters were at a premium and because of my very limited length of service I was a long way down the waiting list. I had turned my efforts to trying to find rented accommodation near Dundas within the limits of the allowances I could claim. I was not having much success when my luck changed quite dramatically. The Adjutant of the 1/15 Royal NSW Lancers, Captain Graham Lovegrove, occupied the top floor of an old house in Lancer Barracks, Parramatta, with the bottom floor being occupied by a Corporal. This arrangement had been very much less than ideal and Graham had been doing the rounds trying to find an officer to take on the bottom floor. I jumped at the offer notwithstanding the quarter being only two bedrooms with a shared laundry and a lean-to garage. The quarter had been left in a pretty dirty condition but I was pleased to have my family with me again.

The 1/15th was an old regiment and very much part of the Citizen Military Forces establishment. The Regiment had only two Centurion tanks in situ that allowed for some limited local training but it seemed to have a full complement of officers and NCOs. It was a moneyed regiment that left me out of my depth on many occasions and I often wondered how the two regular officers (Graham Lovegrove and the Quartermaster, Peter Ryan) coped with their mess bills.

Let me recount one story that left me flabbergasted. I had dropped into the mess on a parade night and overheard two of the Majors, Arnott of biscuit fame and MacArthur-Onslow, questioning the Quartermaster, Captain Peter Ryan, about delays in getting new drums for the band. He explained that the matter was out of his hands but he did not expect to receive the drums for many months. One of the Majors asked how much a set of drums would cost and was advised some hundreds of pounds. The questioner then volunteered to buy the drums himself. The other Major demurred saying he would buy the drums as his companion had bought the last round of drinks!

Peter Ryan was a man of many talents, particularly when it came to the more arcane of the quartermaster arts.

I remember well how Peter used the much loved Conversion Voucher to convert a Nets, Camouflage, Tank worth some two or three hundreds of pounds to a Nets, Camouflage, Helmet worth about five shillings and wrote it off.

The Regiment had a band that used to practise every Tuesday night. This was reasonably bearable while they worked away at learning a suitable repertoire of marches but then, from my point of view, disaster struck. The band was invited to play at the local gala opening of *South Pacific*. For months leading up to this event the band constantly and excruciatingly practised 'Some Enchanted Evening'. The experience put me off musicals for life.

As the year progressed I found that I was getting reasonably comfortable with my appointment but I was not doing too much engineering and there was a distinct lack of intellectual challenge. I decided to seek approval to do a postgraduate degree at the University of New South Wales. I wasn't really interested in a research degree but a mainly coursework Master of Technology (later the name was changed to Master of Engineering Science) was being offered as a two-year part-time or one year full-time degree. The part-time commitment would be half a day plus two nights a week.

My application to undertake the course had a rocky path as my immediate superiors did not see the need for officers with degrees on the basis that all the technical talent that was needed in the Corps could be provided by the Foremen of Signals. Having a bachelor's degree was barely acceptable but a postgraduate degree was thought by some to be over-trained!

I persisted with the application and found an ally in the Chief Signals Officer, Lieutenant Colonel D A C Griffith, mentioned previously. His main interest was in the subjects that I intended to take. When informed these would include Communication Theory, Computers and Advanced Antenna Theory, he became my champion. He then sought and gained approval from the Director of Signals for me to enrol in the Master's programme in 1963.

There was a most amusing incident that occurred towards the end of the year. I was not directly involved but I can assure the reader that the incident did happen and was a source of amusement to all in the Regiment.

Disciplinary problems were somewhat of a rarity and usually handled in a semi-formal way with extra training, etc. However, we had a young soldier who tended to be ever in some sort of minor trouble. After a number of warnings that did not appear to be having the desired effect, a charge was laid. Major Little heard the charge initially but, although the offence was not terribly serious, he referred the matter to the CO to provide a salutary lesson to other would-be miscreants. At the second hearing, held on a Friday, the soldier pleaded guilty and the

CO awarded 14 days Confinement to Barracks.

The following Monday an older, smartly dressed gentleman with distinct military bearing presented himself at Regimental Headquarters, identified himself as the father of the soldier mentioned above and requested an interview with the CO. The visitor was ushered into the CO where he opened his remarks with the question:

"Is it true that you found my son guilty of... ... ...?"

"Yes" said the CO.

"Is it true that you awarded him 14 days CB?"

"Yes" said the CO.

The visitor then slammed his fist on the table almost screaming:

"It is not enough! It is not enough! Have you no discipline? He should be in the cells!"

It transpired that our visitor had been the equivalent of an RSM in the German Army during WWII and held views on discipline very different from those held in our peacetime Army.

There was a silver lining to this episode as it was found our erring soldier need only be threatened with a weekend leave to his father for a magical change in his behaviour to occur.

The year came to an end with my promotion to substantive Captain having now served four years in the rank of Lieutenant.

Major Little had moved from the Regiment in the normal posting cycle and had been replaced by Major Lloyd Solomon. Major Solomon had quite a reputation in the Corps for being a perfectionist and somewhat of a martinet and all members of the Squadron were in some trepidation of his taking up the command. Major Little had been a blusterer whose bark was usually worse that his bite. On the other hand Major Solomon was known to be an extremely hard taskmaster demanding nothing short of perfection in every task undertaken by those under his command. I must admit to some feelings of trepidation on hearing of his posting as Officer Commanding.

Major Solomon made an immediate impression by requiring all officers in the Squadron to read a tract called ' A Message to Garcia' and to sign a form confirming completion of the task. The story extolled the virtue of a young man who completed a formidable task of taking a message from Washington to someone called Garcia in the depths of the South American forests. Allegedly, he sought no clarification of his orders, did not enquire about how he was to complete the task, made no demands for resources but just got on with it. I didn't know then and I don't know now what we were supposed to gain from this example but I can tell you I trod carefully around the Major.

Major Solomon was an elder in the Mormon Church and so had firm views on the evils of smoking, drinking alcohol and the use of caffeine in such beverages as tea and coffee. He also had a distinct tendency to view any dereliction of duty, whether omission or commission, as a moral issue. I had personal experience of this when he paraded me to the CO with the complaint that I was not using my talents and training at the level to which I was capable. It smacked of the old RMC offence of 'Lack of Zeal'. I must have improved as I recall getting on quite well with him after a somewhat shaky start.

Major Solomon was a devout man but he certainly did not attempt to proselytize. He did, however, remonstrate with others and me on how we were killing ourselves with the demons of nicotine, alcohol, tea and coffee. He was doing this after one rather long dining-in night when, possibly because of an ample sufficiency of Dutch Courage, I asked him how he squared himself with being a soldier with the ever present risk of being sent on hazardous duty. He went very quiet and I immediately thought I had made a risky career move.

I must say that I was relieved on the following Monday when he took me aside and advised that he had discussed the issue with a group of church elders who pronounced that he could continue to serve.

Early in the year I presented myself to the University to enrol in the Master of Technology programme. The requirements were Mathematics, three advanced subjects and a sub-thesis. As I had indicated to the CSO, I elected to take Computer Design, Advanced Antenna Theory and Communications Theory. Soon after lectures started we were advised to select a thesis topic for approval by the Dean. I had had a long-term interest in cryptography and thought this would be a topic acceptable to my military masters. How wrong I was!

My initial approach to the military was refused outright due to the security classification of the subject in general. I persisted and was told that I could undertake a project but the report would need to be classified. The University would not accept this and insisted that the completed work would be placed in the library. After further negotiation I finally received a two-page message agreeing that I could undertake the project but listed about thirty provisos detailing what must be excluded. The message ended with the statement:

'You are to confine your research to the wealth of unclassified published material.'

The 'wealth' consisted of one book by Helen Fuch Gaines, which I avidly read.

After reading the book and giving the matter some thought, I determined on a title that I submitted to the Dean. My work was to be called:

'An Analysis of Certain Aspects of Digital Encryption'

The immediate reaction from the Dean was the question 'is encryption a real word or have you just made it up?'.

Having convinced him that it was a real word, my project was approved and I awaited the appointment of a supervisor.

After a week or so I was approached by my supervisor and we had a most interesting conversation skirting my subject and his background. He was obviously ill at ease about the whole matter which I found somewhat intriguing. It transpired that he had been employed by Defence Signals Division before joining the University and the details of his employment were supposed to have been suitably camouflaged yet his first task was closely related to his previous employment. It eventually became clear that his appointment was coincidental and his cover remained intact. I started lectures on two of my subjects and preliminary work on my thesis.

Lectures were held on the Kensington campus of a late afternoon and evening which meant quite a trek from Dundas to Kensington and back to Parramatta. My academic load was relatively mild and I soon came to admire greatly those who undertook a first degree part-time.

Back at the Regiment, life went on much as usual. The CO and Major Solomon seemed to have more that the usual run of turf wars and they developed an interesting means of settling them. I have mentioned before that we were tightly pressed for room so arguments in either office could not be discreetly conducted. When trouble brewed there would be a quick telephone conversation followed by both storming from their respective offices to meet on neutral ground near the unit flagpole. The whole thing had a wisp of comic opera and was a source of great amusement to all ranks.

It was about this time (early 1963) that the Regiment was equipped with new state-of-the-art microwave communications connecting Dundas, Victoria Barracks, Wallgrove and Bringelly. The equipment operated at about 2 GHz and was capable of carrying many multiplexed telegraph channels. It replaced the VHF channelling equipment, not because of capacity requirements but because the VHF frequencies were required for civilian use.

The 2 GHz equipment was very advanced technically and presented many challenges to our older technicians. There were hardly any controls on the tall equipment racks and not much in the way of flashing lights (commonly called OFPs – Officer Fascination Panels) but they did have six dials per rack measuring various voltages and currents. Someone much senior to me decreed that the readings of these meters were to be recorded each 15 minutes and a large leather-bound ledger was provided for the purpose. These records did not assist in fault-finding and, as far as I could see, served no useful purpose whatever. However, it did serve to instil in me a deep-seated suspicion of any type of pro-forma that stayed with me throughout my service.

About the middle of the year I had a call from the CO to inform me that I was to be appointed stocktaking officer for the annual stocktake. This filled me with dread, not because of the complexity of the task but the mind numbing boredom of it.

A good stocktake result was seen as an essential requirement for the

CO's standing in the big scheme of things. I could never understand why the stocktake should take precedence over unit efficiency and morale but this certainly appeared to be the case from my observations. I was also intrigued by the Army's willingness to spend thousands of pounds in time and effort to track down the loss of something worth shillings but I guess it was due to our responsibility to protect expenditure from the public purse.

The stocktake involved counting EVERYTHING on the regiment's books. The stocktake officer would proceed with his entourage of 'Q' staff and recording clerks to count things it would be incredibly hard to move let alone steal (transmitters, receivers, antennae, etc), furniture that could readily be shuffled around and hence had to be marked when counted, soft goods (linen, blankets, etc), kitchen equipment, clothing, tool kits, etc. The counting process took days to complete. This was followed by adding and re-checking columns of figures and comparing the results with the ledger quantities. Discrepancies led to detailed examination of the paperwork backing the numbers in the ledger. At the completion of all this clerical effort, the final result (measured in monetary value), comparing what we could find with what we were supposed to have, was presented to the CO. He either accepted the result or decreed the task be repeated. When all this was done, the final result was forwarded to the next higher headquarters that would either accept the result or order an investigation.

As I said, a mind-numbing process which was incredibly expensive in man-hours and invariably showed that missing items had not been stolen but were the result of faulty paperwork.

I am happy to report that my first stocktake was a success. The QM was elated, the CO pleased and my Annual Confidential Report reflected these feelings.

A POSTING TO THE STAFF

While still basking in the glory of this achievement, I was posted to the staff of the Chief Signals Officer Eastern Command, Lieutenant Colonel D A C Griffith, located in Victoria Barracks, Paddington, as the Staff Officer (Signals) Grade 3. I must say that I was very pleased to move as there was little challenge left for me at the Regiment and my new direct superior would be Major David McMillen, the Staff Officer (Signals) Grade 2.

Moving to the Headquarters brought some unexpected expense. Not only was there substantial travel involved but also I had to expand my civilian wardrobe as wearing uniform was discouraged. Officers were required to wear suits so I went into debt with HG Newton and Son once more.

I was given a week to settle in to my new surroundings and meet other members of the Eastern Command Headquarters staff which was quite substantial and commanded by a Major General. In those days the Army was dominated by the Eastern Command (in NSW) and the Southern Command (in Victoria). Both were commanded by Major Generals, both appeared to show great disdain for Army Headquarters and there was fierce rivalry between them. I should also mention there was also the overlay of the Citizen Military Forces that wielded great influence if not power.

About mid-morning of the first day of my second week, I was called into the CSO's office to commence my training as a staff officer. The CSO sat at his large desk with the SO (Sigs) 2 at his right hand and me at his left equipped with note book and pencil. The routine was simple, the SO2 passed correspondence to the CSO, the CSO scanned it and generally (particularly if the letter came from the Directorate of Signals) consigned the material to the wastepaper basket but occasionally remarking to the SO2:

"this seems a good idea, wait for a week or so and then write to AHQ as if we thought of it first."

I expect that the whole performance was put on for my benefit, but

I certainly learnt that one should not automatically react to letters from on high!

The CSO was a quite remarkable staff officer. He could cut to the crux of an issue in short order and then quickly produce cogent, well-argued solutions in beautiful prose.

He was a hard taskmaster but always explained why he rejected all or parts of drafts put to him. He was also fiercely protective of his staff and was one of a rare breed that took on to himself any criticism of work emanating from his office while ensuring any praise was properly directed to the individual responsible for authorship.

He also had an aversion for people wishing to appear busy when there was nothing on. If things were slack we were sent off to do something useful like sport, etc while expecting 150% when we were under pressure. He really was a joy to work for.

I never did get used to him appearing in my office towards the end of a working day and asking in his well modulated and cultured drawl (which I learned to imitate quite well):

"Oh Peter, will you be in tomorrow?"

On receiving an affirmative reply, he would give me a task to complete.

I had been an Honorary Member of the Lancers' Mess for about two years until one fateful Saturday morning in late 1963. I was looking over the back gate waiting to see the QM, Peter Ryan, when a brand new Second Lieutenant appeared. He was resplendent in a black tanksuit (much like a standard boiler suit but with more pockets), a white silk cravat, a black beret with rooster plume, a Sam Browne Belt with sabre and he carried a cane. I'm afraid the sight was too much for me and I mumbled some comment about 'weekend warriors'. I thought the comment had gone unheard but I was wrong. Within a matter of an hour or two, I received a formal note cancelling my Honorary Membership of the Mess. So now I had been rejected by two Corps – Artillery and Armour! Fortunately, the Corps of Signals was less choosy.

I remained on the CSO staff into 1964 and received a very complimentary Confidential Report for my efforts. He couldn't resist injecting some humour as he added a remark against the first item on Appearance (the highest grade being Impressive with Good Bearing) with the remark:

'Would be impressive if six inches taller'.

SPECIAL FORCES

The CSO having decided that I had received a suitable grounding in staff work, posted me to an ARA/CMF integrated unit, 301 Signal Squadron, based at Lidcombe which was quite close to my married quarter at Parramatta. I was to be the ARA Troop Commander, Training Officer and Squadron Operations Officer.

I shall never forget the interview I had with the Officer Commanding the squadron when I presented myself early one Monday morning. After the niceties on past appointments, etc had been exchanged I admitted complete ignorance of the roles and functions of the squadron other than it was an integrated unit. My request for information was rejected with the admonition:

"*You are not [security] cleared to know*" quickly followed by:

"*You are to attend a briefing at Victoria Barracks tomorrow after which all will be revealed.*"

Having gathered my wits, I asked if I could see my troop equipment. The OC then called for the Quartermaster who took me into his storeroom in which was a large vault. Inside the vault I was shown a number of leather cases, looking for all the world like small luggage, in which were housed compact radio sets. Having seen something similar in many WWII movies I realized that I had become a spy!

As mentioned above the unit was an integrated one with a small troop of ARA soldiers and a much larger number of male and female CMF officers and soldiers. The CMF component was extremely keen and, for the most part, very competent. The training standards were high with 25 words per minute Morse Code being the basic requirement for membership of the base station troop. To achieve this standard there was a continuous Morse course run one night per week in addition to the normal weekly parade and the monthly camps. This level of activity was much in excess of the available paid time so the dedication of the CMF members was patently obvious.

Not all the officers were as good as the soldiers under their command. One in particular had been a probationary lieutenant for over ten years. I attempted to sack this man and had the whole weight of the Imperial Service Club descend upon me!

Not long after joining the Squadron I found that I had become a 'volunteer' for parachute training. I was pleased at the prospect and made quite an effort to improve my level of physical fitness before proceeding to RAAF Williamtown for the four-week course.

Soon after my nomination for parachute training became known, I was invited into the Sergeants Mess at Dundas where two old friends, Dudley Callaghan and Swede Larson, decided to commence my training by demonstrating exit techniques from the aircraft by jumping off the bar. Swede became very enthusiastic and pushed off with so much vigour that his head penetrated the ceiling where he hung for a short time before falling to the floor with less hair than he had when he started.

There were four officers on the course and probably twenty NCOs and soldiers. The psychology of the course was interesting. When something frightening (jumping from the Polish Tower for example) had to be done, the officers were detailed to go first. We, of course, did not want to flinch in front of the soldiers and, in return, they were keen to show that they could do anything an officer could do. The approach seemed to work. I did not do well and did not survive the ground training and was returned to my unit. I did eventually get to jump out of an aircraft some twenty years later when I did a water jump.

I very much enjoyed my posting to 301. We had limited support and did a lot of self-help projects including design and fabrication of a mobile base station. This was particularly successful and much before its time. These projects required considerable quantities of industrial gases that we obtained through the normal supply system. The system usually worked well but there was the occasional hiccup as the following examples show.

I indented for a large amount of oxygen gas to complete a project. After a number of weeks and many phone calls some empty gas cylinders were delivered with the Indent annotated 'Oxygen to follow'!

The second little difficulty I had concerned an attempt by me to acquire the newly introduced into service 45 mm automatic pistols. My Indent was 'satisfied' by the system delivering Smith and Wesson revolvers with magazines for the pistols!

While all this was going on, I completed my thesis for the Master's degree and sat my final examinations. Some little time before Christmas I was advised that I had met the requirements and would be admitted to the degree in early 1965.

Shortly after this, I was summoned to Canberra for an interview with the Director of Signals. I was expecting a big pat on the back and a good posting and so went to Canberra with great expectations.

I arrived at the Directorate and was duly summoned into the presence of the Director himself. I must say that I was feeling rather good about my achievement and was expecting suitable reward. I was invited to sit and then the Director said:

"Now Peter, I want you to listen carefully to what I am about to say. This Master's degree of yours – it won't be used against you in any way!"

As my jaw dropped to the floor he added:

" I have a posting for you that will use all your new knowledge on Antenna and Communication Theory – you are to be Station Commander Diggers Rest."

I felt pretty deflated as I left the Director's office, particularly as I had already spent time at Diggers Rest installing transmitters in the new building. I then visited the Lieutenant Colonel in charge of RASigs postings and soon realized that what the Director had said was that I would not be regarded as a boffin but would follow a normal career path. He had also given me a substantial command for a Captain so I returned to Sydney in a good frame of mind.

BACK TO DIGGERS REST

The year started with preparing inventories for the move to Diggers Rest. This was always a time-consuming and boring task that invariably led to arguments between husband and wife about what to take and what to throw away. I was to move into a brand new married quarter at Diggers Rest but as completion of the house was some way off, I took up residence in the Officers Mess, Watsonia Barracks and commuted each day to the Station in my chauffeur-driven Holden utility!

The Transmitting Station had been completed and was a major link in the Australian Communications Army Network with circuits to all Australian capitals and into the US, UK and NZ networks. The station operated 24 hours a day seven days a week so I had the usual problems that came with shift workers. I was the only officer on establishment although occasionally, Second Lieutenants were attached for training.

The other outstation was the Receiving Station, Rockbank, commanded by one of the Corps legends, Captain 'Happy' Haines. 'Happy' had sustained serious injuries as a Despatch Rider and spent much of his time in pain and was often grumpy – hence the nickname 'Happy'. I found him most helpful and I consulted him often on administrative and man-management issues.

Lieutenant Colonel 'Lauchy' McLean commanded the Regiment and my immediate superior was Major Reg Topp. The latter was an interesting man whose actual rank was Captain but had been a temporary Major since WWII and had remained so due to his refusal to sit for promotion examinations. Both would visit me from time to time but generally speaking I was left alone.

Not long after arriving at the Station, I travelled back to Sydney for the conferring of degrees. I had not been able to do much by way of preparation for the big event and wasn't able to arrange for the hiring of a Master's gown and suitable hood until the big day. Luckily, I went to Kensington early and discovered that the only gown available was to fit someone 1.8m tall or over. I had to make a mad dash to the tailor at Victoria Barracks to have the gown temporarily shortened by 30 cm.

The Conferring Ceremony was a grand event and I must say that it was a proud moment for my family, my parents and me when I was admitted to the degree.

We returned to Diggers Rest and took up residence in the OC's house. The house was pretty grand by the standards of the day with three bedrooms, a good kitchen, dining room and lounge plus a carport and a concrete driveway.

There were about ten other married quarters on the station of varying standards, a Sergeants Mess, an Other Ranks Mess, a Q Store, Vehicle Compound but not much else by way of amenities for the families. The 'town' of Diggers Rest, located some thirty kilometres from Melbourne on the Calder Highway, consisted of a railway station, Post Office, a store or two, a small primary school, a garage and a hotel to serve a few small farms in the region. The nearest town was Sunbury which was of considerable size and a centre of horse studs, farms, etc. It was also the location of the only doctor in the region.

One of my major concerns at Diggers Rest was the prevalence of tiger snakes and the possibility of someone being bitten. As the Sunbury doctor was the only one covering the region, he was unwilling to hold or administer antivenom. This was understandable but it left the prospect of a child being bitten with the nearest help being the Children's Hospital in Melbourne. I'm pleased to say that no such event occurred during my time with my personal experience of tiger snakes being limited to being bailed up by one as I walked to work.

Not long after arriving at Diggers Rest I incurred the wrath of the local shop owners by turning on a weekly shopping bus to Essendon. This was certainly offset by the gratitude of the married patch residents who had suffered high prices and lack of choice at the local stores.

There was not much in the way of technical challenges for me as OC but I had plenty to occupy myself with administering the station and the staff. There were also many visitors who were interested in inspecting a 'state of the art' transmitting station. At one stage I thought I would make a career out of guiding visiting Indian and Pakistani officers around the outstations due to the large number of visitors. I found the escort duties demanding as I was forever being tempted to mimic Peter Sellers when answering questions!

Life on the station was not easy for my wife, Yvonne. Diggers Rest, although only twenty miles from Melbourne, was isolated and she had very little opportunity to make friends outside the station or to find intellectual stimulation. If she took an interest in what the other wives were doing she was 'interfering'; if she didn't, she was open to criticism for being 'stuck up'. I suspect my children may also have been subjected to some harassment in their small school.

There were a number of characters on the Station staff that are worthy of mention. Firstly, there was WO1 Bill Dickenson who was my Foreman of Signals. Bill was a very competent senior, an absolute gentlemen and a mentor as I built on the skills and knowledge gained at RMC and university. Bill became a firm friend and I valued his friendship throughout my career. On a very different plane was SSgt Jack Wicks who later became the Mayor of Sunbury. I think it would not be unreasonable to say that Jack was my Sgt Bilko!

Then there was Sig Russell .... who was the General Duties soldier (odd jobs man) on the establishment. Russell was not the brightest of soldiers and needed pretty constant supervision. For example, if Russell was asked to dig a hole someone had to be on hand to tell him when to stop digging. He might have been slow of speech and intellect but Russell was completely honest and trustworthy. I used this to advantage when I put him in charge of the Other Ranks' Canteen. Service may have been slow but I never had any further worries about unbalanced books, missing stock, etc.

Russell occupied a Married Quarter with his youngish wife. I discovered that this young lady had a very healthy libido as every few weeks Russell would come to me asking to be put on weekend duty that would require him to 'live-in' so he could get some rest. In explanation, Russell told me that every day on returning home from work his wife would say something along the lines of 'Russell, come into the bedroom, I want to show you something'. This would be followed by some hours of hectic activity that left him exhausted.

I also had to deal with a local farmer who made a habit of using his tractor to break the water pipeline from Sunbury to the base, waiting until his land was suitably watered and then notifying me of the break and seeking compensation for damage.

Towards the end of 1965 I was informed that I had been identified as a candidate for a posting to the USA to attend training in satellite communications. I subsequently was interviewed by the Director of Signals and told I had been selected. I could not be given a date of posting as negotiations were still proceeding with a number of US Agencies but I should expect to leave in about six months. I understood that the major stumbling block was the highly classified nature of US satellite systems. There then commenced an on-again off-again cycle that continued until I finally left for the US in April 1967.

Early in 1966 I discovered that one of my administrative staff had been stealing small amounts of money from fares collected for the wives shopping trips. I charged the soldier, the only time this occurred throughout my service, and he was subsequently dealt with by court-martial.

The on-again off-again saga of the USA posting continued throughout the year and I was put on a three month notice to move. I had been called to Canberra a number of times to discuss the proposed schooling, as there was some doubt that I would be able to handle the course. These doubts were caused by the course description that had been obtained from the USA Signal School, Fort Monmouth. I undertook pre-course study and spent a lot of time working on orbital mechanics, microwave systems and anything else that I thought would be relevant.

During the year I completed my examinations for promotion to major and attended the appropriate courses at the School of Signals and Canungra. Under the normal scheme of things I could expect to receive temporary promotion at the end of the year with substantive promotion to occur ten years after graduation from RMC, i.e. in 1968. Given that I expected to serve part of the next few years away, I thought this was my most prudent course of action.

Members of RASigs were being posted to Vietnam but not in large numbers. At this stage the posting policy appeared to be to start at the top of the seniority list and work down. This meant that I was unlikely to be selected until return from the USA.

Life was pretty quiet at Diggers Rest apart from the occasional pushing match in the Other Ranks Canteen and domestic disturbance in the Married Quarters. Social life was basically limited to mixed functions at the Officers Mess at Watsonia.

WATSONIA

After almost two years in command at Diggers Rest, I was posted to the Directorate of Signals in Canberra but detached back to 6 Signal Regiment. The posting involved promotion to Temporary Major and a move into the Married Quarters at Watsonia.

The saga of the overseas training continued with the notice to move reduced to two months. My promotion to Temporary Major caused a hiccup in the USA as the original negotiations had been for a Captain. This problem was resolved and we all went to Sydney for Christmas.

Early in the year I received an urgent call to go to Canberra to attend a selection board for attendance at the Royal Military College of Science in the UK. I must admit to a degree of excitement as I was much more interested in travelling to the UK than the USA and the posting would be for two years rather than the one year for the latter.

The interview went well and by the end of the day I had been told that I had been selected. What I hadn't been told was that no one in the Military Secretary's office had bothered to tell the Director of Signals about the proposed posting. I called on the Directorate after the interview to inform them of my success. This was the first that Officer Postings had heard of the matter and I was accused of having applied for the course without the knowledge of the Director. I finally convinced him that the call to Canberra for an interview had come out of the blue and was not instigated by me. While I was flying back to Melbourne there was an almighty row going on between the Military Secretary and DSigs which the latter won. I arrived back in the Mess at Watsonia to be told that my selection for RMC of S had been withdrawn so setting some sort of record for the shortest-lived overseas posting. I settled back into work at 6 Signal Regiment that was now under the command of Lieutenant Colonel Ken Hill, an absolutely delightful man who became a great friend.

In March I received the long awaited Posting Order, sending me to the USA Signal School for six months of courses to be followed by a six-month attachment to a satellite ground station. Inventories were prepared, luggage bought, plans made for stopovers and our car put up for sale. All went well until a week before departure (a Friday) when the Posting Order was cancelled once again.

I stopped the car sale and settled in for lots of recriminations from my family about the Service.

On the following Monday the Posting Order was reinstated. Unfortunately, the signal arrived too late for me to conclude the car sale that I now had to sell to a dealer at a substantial loss. However, we proceeded to fly to Sydney for a few days with parents before starting the big adventure. We attended a few farewell parties and then finally went to Mascot for a 20:00 departure on a QANTAS 707.

After tearful farewells from mother and mother-in-law we boarded the plane. Just before we were about to push-back from the gate a steward paged me to advise that an urgent message had arrived for me. I expected the worst and was mentally preparing to disembark when I found, with great relief, that it was only a farewell note from friends who had missed us in the Departure Lounge!

OFF TO THE USA

These were the days when officers travelled First Class irrespective of the means of transport. The normal means of overseas travel was by sea but time did not allow this so it was QANTAS to San Francisco via Hawaii and thence to Washington, DC, for briefings.

The flight was uneventful but long and we eventually arrived at San Francisco where we went through arrival formalities. Because we were on a twelve month posting and shipping would be too long, we had been given approval for a considerable amount of excess baggage to allow us to bring household linen, etc with us. This amounted to three trunks and nine suitcases together with hand luggage. Immigration was no problem but, unfortunately, we were confronted with a black, female and very officious Customs officer who insisted that I open every piece of luggage for inspection. This stretched my patience and that of the family almost to the limit as we had been travelling for over twenty-four hours.

Eventually we left the airport in two taxis to go to our overnight accommodation. I had been briefed on the local tipping conventions but I was ill-prepared to tip the cabbie, the porter who took our baggage to the hotel reception and another porter who took us to our rooms.

Our next experience of local culture was at the hotel dining room. We placed our orders and almost immediately salads, that we had not ordered, were placed before us. Nothing more happened so I called the waiter to enquire about our meals. It was then made clear that nothing further would happen until the salads had been consumed!

The next step was another long flight to Washington, DC for briefings. We had a day or two there and then took the short flight to Newark, New Jersey. The Allegheny Airlines flight, although short, verged on the terrifying and we were grateful to step onto terra firma. Not long after our arrival, Allegheny Airlines lost a stewardess out of an improperly closed door confirming the opinion we had formed.

On arrival at Newark, I arranged for two cabs to take us to Fort Monmouth. This was not as simple as one might think as the journey was about 100 Km and involved two toll roads. The initial part of the journey was through the industrial area of Newark with oil refineries and lots of chimneys belching black smoke. I was cursing myself for bringing my family to what seemed a godforsaken wasteland when we entered the aptly named Garden State Tollway with scenery to match – what a relief?

At Fort Monmouth we were assigned rooms at the Post Guest House and I started my briefings with the Foreign Liaison Office including an English test.

The Foreign Liaison Staff were efficient, friendly and helpful and they soon put me in touch with a local civilian employee who was departing for Germany on posting. We quickly settled on rental arrangements and the family moved to Sea Girt on the Jersey shore some 50 km from the post. The owners were a delightful couple who did everything to help us settle and topped this off by giving us a very large Chevrolet station wagon. The vehicle was enormous and guzzled petrol at the rate of 8 miles per gallon. However, petrol was cheap and the vehicle was a useful addition to the Rambler I had purchased for just $900.

The house was brick (unusual for the area), very large and on a sizeable block of land. In short, far superior to anything I had ever occupied. One condition of the lease was that we continued our membership of the Beach Club including the beach hut. This was because the beach at Sea Girt was private and only accessible to club members.

After a week of settling-in, I started my course on satellite communications at the School of Signals. I must say that I had been very nervous about the course and concerned that I would not be able to keep up. Imagine my surprise when I found that the course was mainly aimed at technicians with little required in the way of prior knowledge. I had no trouble keeping up and generally topped all the modules with a minimum of effort.

At the end of each module, there was a mini-graduation with certificates distributed by the Chief Instructor. At the conclusion of a particular module on a spread-spectrum modulator we had the usual ceremony but my name was not called. I queried this and was told, very politely, that I had not attended the course. Patently this was untrue and I said so. In a voice that was a little louder I was again told that

I had NOT attended the course. Before I could say anymore, I was taken aside and told that the course was security classified at a level that precluded foreigners from attending so I then quickly agreed that I had indeed not attended.

Life at Fort Monmouth was very pleasant. My studies were undemanding and the overseas allowances meant that there was ample money to enjoy our stay. We had access to the PX and Commissary plus an adequate allowance of near duty free liquor for entertaining. So entertain we did in our very pleasant and comfortable Sea Girt house.

We also regularly attended Mass at the Post Chapel followed by brunch in the Officers' Club. It was at one of these functions that I met the Commanding General, Major General Thomas M Reinzi. The good General had met many Australians in Vietnam and had become very fond of Fosters Lager which I managed to get for him from time to time.

The course took about six months and I was scheduled to spend the next six months on exchange at a satellite communications station at Fort Dix. I duly arrived and was warmly greeted by the Station Commander who was also a Major. I was allocated an office and settled down to being thoroughly schooled on the operation of the station.

On arrival at the station on my second day, I found that some wag had placed a sign over my office which read 'Commanding Officer Australian Forces, Fort Dix'. This title stuck and led to some embarrassment at a later date.

It soon became apparent that it would not take long for me to learn everything I could learn from the station so the CO took me to meet the Commanding Officer and senior staff at the Satellite Communication Agency back at Fort Monmouth. I advised the Army Staff at our Embassy of this and a second line of approach was taken by them to get my posting changed. This was ultimately successful and it was agreed that I would move to the Agency in early 1968.

While a student at Fort Monmouth, I made friends with officers of the Foreign Liaison Office and also with a Major (later Lieutenant Colonel) James E McKeon. He had already completed a tour of duty in

SVN and had met a number of friends of mine. We became firm friends and I was able to keep in touch with him and his wife Carol for many years.

As the year wore on Jim asked me about my Christmas plans. I had none so Jim asked my family to join his in his 'shack' in New Hampshire. I must say I was rather diffident about spending Christmas in a shack and put him off a few times. Eventually, we decided that, as we had nothing else to do and as we were expecting to return to Australia in April 1968, we might as well take up the invitation.

Christmas came and I bundled the family into our Rambler and headed north. The trip was slow as I was nervous about driving in the snowy conditions. As dark fell we arrived at the 'shack' which turned out to be a luxurious house on the banks of Lake Winnipesaukee. After a day at the house, skating on the lake and generally having a lovely time, Jim announced that we had all been invited to join Uncle Carl for Christmas lunch. Another revelation followed – Jim's uncle was apparently one of the wealthiest men in Boston.

Back at the Lake, Jim asked if I had ever seen a timber logging operation. I confessed I hadn't so Jim suggested both families travel north to view one. We went up almost to the Canadian border where we booked into a very comfortable motel. While wives and children stayed at the motel, Jim and I took off and I was shown extensive timber holdings and a large saw mill. After a pleasant two night stay we headed back to the Lake.

When I went to pay the motel bill, I was informed we were guests of the house. I was very surprised at this so after getting back to New Jersey, we sent gifts to the motel manager and his wife. I was later to learn that Jim owned the motel and the logging operation! Jim was, in fact, a very wealthy man who just liked soldiering. Many years later when he retired he bought a bank which will give you an indication of his resources.

Some time later Jim revealed the source of the family wealth. Jim's great-aunt apparently ran the best 'house' in Boston. This was a great source of shame for much of the McKeon family but not for Jim's father who regularly had Aunty Kate to the house. Ultimately this resulted in a considerable fortune coming to the family who, with very astute management, turned it into a much larger one.

Further gold entered the McKeon coffers by way of a bachelor uncle who was one of Boston's finest. Apparently he managed to get permanent traffic duty on building sites where a dollar bill held in the extended hand of construction site truck drivers ensured that traffic was stopped to facilitate truck movements.

I must say that I was mightily impressed with the McKeon wealth but even more impressed with their reluctance to display it.

Just prior to the end of 1967, I was informed of the traditional requirement for all officers on the base to attend on the Commanding General in the Officers Club on New Years Day to pay respects. I gathered this was a long-standing tradition where officers donned their best uniforms and paid their respects to their General. This lead me on a flurry of activity to locate and polish my Sam Browne Belt and other bits and pieces to dress up my 'Blues' uniform for the big day.

It was quite a grand affair with all attending joining a receiving line to be presented. Fort Dix is an infantry base so the only drink served on the day was the blue infantry punch! Yvonne and I were presented to the General who was more than mildly surprised to find he had an Australian Signals officer on his base. This was my first and last time in the Fort Dix Officers Club as the satellite station officers generally used the much closer Lakehurst Naval Air Base Station Mess.

The move to SATCOMA was extremely beneficial to the Australian Army, as it had been lobbying for some years to get access to satellite technology. What had been unsuccessful by long-range negotiation was overcome by local efforts. These local efforts were very much helped by the fact that the Commanding Officer of SATCOMA, Colonel Jerry Rippey, had met a number of Australian Officers in his previous postings, notably an old friend of mine Bob Cowley, and was very much predisposed to have an Australian attached. What added to his enthusiasm was the fact that he was short of qualified engineers and I came to him at no cost.

Initially, I was to be a Liaison Officer but I quickly palled at this idea and I approached the CO and asked that I be given wider responsibilities. His response was "I was just waiting for you to ask" and by the end of the day I was project manager for an asynchronous Multiplexer, one of the high priority projects of the Agency.

I had the best part of a year working at the Agency in a way that would not have been possible at that time in Australia. I had a budget to manage and a reasonably free hand. I had access to the latest technical developments and enjoyed the company of military and civilian engineers who were at the forefront of development of satellite communications.

This was a very exciting time as the technology was still very much in the early stages of development. The programme was known as the Initial Defence Communications System Programme. Launch vehicles were not yet powerful enough to place satellites in synchronous orbit so the approach was to place small satellites in clusters in just sub-synchronous orbits. This meant that ground stations had to have the ability to track satellites and it was necessary to have elaborate procedures to hand-over from one satellite to another as they came into and out of view. The received energy levels were extremely low and receivers had to be cryogenically cooled to reduce receiver noise. Needless to say that this requirement together with the need for tracking antennae made transportable terminals a difficult proposition. Towards the end of my time with the Agency, a developmental terminal appeared with a cloverleaf antenna which augured well for mobile terminals for tactical use.

One of my colleagues was Bob Langelier, a conscript Captain with a PhD. Bob was tremendous value to the Agency and played a major role in establishing satellite communications technology. Bob was an expert on communication coding and made a major contribution to the field by implementing decoders that had only existed in theory at that time. A few years after leaving the USA, I was very sorry to hear that he died young of cancer probably related to high energy microwave radiation.

There was another conscript Captain on the staff at the time who was an entirely different kettle of fish. I had an altercation with him over a testing programme involving a West Coast terminal. He became very insubordinate and I warned him to desist or face the consequences. He told me to do what I liked so I went to the Chief of Staff (COS) and requested an interview with the CO for me to parade the Captain. This caused great confusion with the COS who thought I wanted a band to mount some sort of parade. When I explained that what I wanted to do was to make a complaint to the CO in the presence of the Captain, the COS asked me to commit the issue to writing. I was reluctant to do this as I thought the issue was relatively minor although it could not be ignored.

I never did get to parade the Captain. He was dealt with by the CO under Article 15 and fined one month's pay. I was not even called to make a statement so I learnt just how vicious the US Army system of justice could be. I found out later that this event was the proverbial last straw as far as the Captain was concerned. He remained at the Agency and did good work. He also became one of the politest Captains I have met!

As the year progressed, I became involved in a programme to test the effects of jamming on the Initial Defence Communications System Programme. The jamming signals were provided by a high powered radar and the effects examined by observing error rates on a data stream. The programme was highly classified and reports were marked Secret NOFORN so I was not authorised to read the reports I had written as soon as they were converted to print. I am sure I was not the only exchange officer to have a similar experience.

As this programme progressed I was asked to lead a delegation to the UK to discuss our project and future collaboration. I cannot remember much about the trip except for three things that had little or nothing to do with the project.

The first relates to a visit I made to Moss Bros to buy some khaki shirts. I had come to the USA on a twelve month posting and was now well into my second year and rapidly needing some uniform replacements. The USA shirt was readily available but of an unsuitable colour and Australian Army Staff, London, suggested Moss Bros or Gieves and Hawke. Moss Bros was more convenient so I made a foray there. I was greeted by an overdressed salesman with an attitude that was simultaneously subservient and condescending. I asked to see some khaki shirts and received the response 'What regiment, Sir?' He then proceeded to show me a bewildering variety of colour (green to the lightest of light fawn), fabric (cotton through to silk) and texture (hairy to smooth). The closest was a van Heusen cotton with French cuffs which I purchased. I then asked if there were Australian Army ties in stock, only to be told in a most offensive way that Moss Bros did not stock the ties of the colonial forces!

The second event shows in no uncertain way how innocent I was of the ways of the world. One of the UK companies we visited took our party out to dinner and a show at the Top of the Town. At the end of the dinner our hosts indicated that their hospitality now ended and suggested that, if we wanted to carry on, we should visit a nearby nightclub called Churchill's. We took the advice and took a cab to Churchill's where we paid a substantial cover charge before being shown to a table. Drinks were expensive but not exorbitant and we watched a first class floorshow. When this was over we looked around and saw there were a lot of unattached females sitting around the edge of the dance floor. I was deputised to ask one to dance. When I made my approach, the young lady said I would need to check with the manager. I thought this might be something like laws in the Southern States which regulated public morality. I duly approached the manager who readily agreed to my request but said I was expected to take the young lady back to my table.

We started to dance and the conversation went along these lines:
"First time in the Club?"
"Yes, it's my first time to visit the UK."
"Do you realise I'm a hostess?"
"That's interesting, what airline?"
Interspersed with giggling – *"No, I'm a club hostess."*
"What does that mean?"

"You pay me £20 to sit at your table, I only drink champagne and you buy my cigarettes."

"What, £20 just to sit with us?"

"Yes indeed."

With that she very graciously said, *"I thought you may not know, let's finish the dance and you can return me to my place."*

With much relief I did just that and soon after we left the Club.

The third event was much less exciting but amusing. I travelled separately from my colleagues as under Australian rules I travelled First Class. For the return trip, I was seated up the front on a Pan Am jet at the end of the runway. As we started the roll, the pilot came on the intercomm and told us the approximate flying time and the weather at destination. He then said:

"Now why don't you all lean back and relax – there's no point all of us being nervous!"

I returned to SATCOMA and continued working in the systems area, with particular responsibilities towards the jamming studies. Around this time, my direct supervisor, Mr Leo Lebanca, asked me to accompany him on a visit to the Pentagon. Prior to the meetings that he had arranged, he decided to visit the main Pentagon communications facility to check progress on arrangements for imagery transmitted by satellite for presidential briefings.

We got to the basement and Leo said that I would have to wait outside for him as entry to the communications area was restricted to high level US security clearance. He was about to leave me when he said something like 'Damn it, let's try to get you in'. It ended up being surprisingly easy. Leo marched up to the guard and said 'This is Major Evans, an Australian officer on my staff and I want to show him some equipment'. The immediate response was 'Sure thing – come inside'!

While I had been working at SATCOMA the MALLARD Project had progressed to the point where a system evaluation was to be performed. The evaluation required representatives from the participating countries, and I was detached from SATCOMA to assist. The visitors from Australia were my old friends Lieutenant Colonels

David McMillen and John Bird who were on a three months temporary duty assignment.

The total evaluating team was quite large and the US took over a hotel in Asbury Park for office accommodation. Substantial money was spent on security arrangements with an elaborate system of passes and tokens to control access.

After some initial general work on evaluation, I was assigned to what was called the Cost-Effectiveness Factor Group headed by a British civilian. After a couple of weeks' work in the US, the whole group was transplanted to the UK for three weeks or so to cost the UK contribution to the system. I was lucky enough to be able to bring the family over for part of that time and we were able to visit my old friend Jim Messini at Blandford, where he was doing a Long Telecommunications Course. We were also able to take a quick weekend trip to Paris.

The evaluation work extended into the New Year and I was formally reposted to the MALLARD Office.

MALLARD

My involvement to date with MALLARD had been limited to my part in the evaluation process. I now had the opportunity to delve more deeply into the project as I had access to all the documentation.

MALLARD had evolved as a concept aimed at providing an area communications system not tied to the chain-of-command that would provide totally secure speech, facsimile, telegraph and data communications. There would be a backbone of interconnected radio relay systems with an overlay of mobile access through radio access terminals. The overriding requirement was for high quality secure speech providing speaker recognition. This requirement was not seriously challenged and its acceptance ruled out vocoder based secure telephone systems.

Secure facsimile was an evolving requirement and many thought that this mode of communications would supplant the normal 75 bps telegraph system with enormous savings in manpower. In fact, some protagonists claimed that the manpower savings resulting from the provision of secure facsimile would pay for the whole Project. Notwithstanding the promise of facsimile, low speed telegraph and data were included in the proposed system.

A MALLARD Office had been established in 1967 in a leased building with staffing from the four ABCA nations in proportion to each nation's commitment (from memory Australia was 3%). I had had little to do with the project until 1968 except to attend some meetings at the Australian Principal's house. The Principal was Lieutenant Colonel Les Moore and he had assistance from one or two Australian engineers, one of whom was Rex Christensen who I met much later in my career. Major Barry Tinkler had arrived in Fort Monmouth in mid-1968 to do a long communications course at the USA Signal School and he was also involved in these informal meetings. A further addition to the Australian team was Lieutenant Colonel (later Major General) Cliff St J Griffiths who arrived in June 1968 to work in the User Requirements Section for MALLARD.

There was a minor drama in early 1969 when it was discovered that the MALLARD building was actually owned by the Mafia. There was a great flurry of activity with sweeps for bugs, etc but nothing was found. There was also a rather humorous storm in a teacup at about the same time regarding the height and spacing of flagpoles – what was the order of precedence; should the major players have higher poles, etc all at a time when the project was under serious challenge.

Towards the end of 1968 David McMillen replaced Les Moore as the Australian Principal of the MALLARD Project. While David and his family were in transit accommodation near Fort Monmouth, we had them down for the longest barbecue ever. In the midst of a very pleasant afternoon, snow started to fall. Nothing major had been forecast, so we all enjoyed the cavorting in the dry snow. The snow continued to fall and it soon became apparent that a return trip to Fort Monmouth would not be too easy. In fact, we were snowed in for two days and our household expanded to eight souls.

We had another Mafia related incident in early 1969. Cliff Griffiths' wife Erica had decided to attend a Mafia funeral in Asbury Park. She was arrested by the FBI on the basis that she was the only one there that they did not recognise! Presentation of Official Passports meant little and it took quite some time for Erica to be returned to the bosom of her family.

Work continued on the MALLARD Project but there was growing disquiet among the junior partners (Australia and Canada) on the distribution of the manufacturing work that would flow from the Project. It seemed that Australia's share (3%) of the work would be at the very low end and would be limited to a minor switchboard and equipment peripherals such as radio handsets. This would mean that there would be very little in the way of technology transfer which could offset Australia's costs. The Canadians were in a similar position and they shared our concerns. There were also major difficulties arising between the main players, the US and the UK on the distribution of manufacturing contracts but much of this was hidden from the minor players.

MALLARD was a billion dollar project. The stakes were high for industry and Congress was becoming more and more interested in manufacturing contracts that would go offshore. There were some US interests that would prefer the Project to be scrapped rather than lose major manufacturing contracts.

It was not only US industry that was pressuring Congress but there was also disquiet among the other US Services that the Army was getting more than their fair share of the defence dollar. There seemed to me to be a concerted effort by the US Marine Corps and the US Air Force to expand system technical requirements to the point that the Project became unaffordable.

From an Australian point of view, another difficulty that was emerging was the scope of the Project. MALLARD was an area communications system and, as it developed, it became more and more likely that the system would be based on Corps level operations with limited systems operation below Division. Thus the applicability to

Australia would diminish as our interests were at Division and below. So what had promised to be a great step forward in capability and manufacturing development was looking less and less attractive.

As they say in the US, most of what was happening was above my pay scale and I had enough to do with the work that was proceeding on refinement of the requirement documents. The work was demanding and fascinating and provided me with invaluable experience in equipment development and project management. I had also made a number of friends in the US, Canadian and UK Services, both military and civilian, whom I would encounter again and again in my future career.

Apart from all the above, living in the US for more than two years had made a marked impact on our financial situation. The exchange rate was about $1. 25 US to the dollar and what with allowances and access to the PX, etc, we had been able to save some money notwithstanding a fair amount of travel.

The extra money had also allowed me to indulge my interest in cars. I had bought a two year old Nash Rambler for about $900 soon after arriving in Fort Monmouth as a family car. I had been given a big Chevrolet Station Wagon by the owners of the house we rented but this soon became a liability as it was unreliable and a gas guzzler. I was able to indulge myself with a sports car – first a Triumph TR3 quickly followed by a VW Karmann Ghia convertible. The latter was not really a sports car but it was great fun to drive. I was very sorry in later years that I had not brought it back to Australia.

After a little over two years, it was now time to go home. I received a Posting Order to be an engineering staff officer in the Directorate of Signals in Army Headquarters, Canberra. This was not unexpected and I anticipated two years in that posting.

In those days the normal means of travel was by ship and arrangements were made for us to travel by air to Los Angeles to join the SS *Orsova* for the trip home. Our intention was to arrive in LA with plenty of time to visit Disneyland, etc before joining the ship. We had commenced the removal process, had completed packing and had moved into a local motel for our last week or so.

Phil Clover had arrived in Fort Monmouth to attend the same course that Barry Tinkler had just completed. The Clovers moved into our house in Sea Girt and had bought our Rambler car while the Tinklers had already travelled West where they were to join us on the SS *Orsova*.

We were a few days short of leaving Sea Girt when there was a minor incident in the local school playground that led to a child falling over and 'possibly' being concussed. Within a matter of hours the family of the child was talking of litigation directed towards me even though the school had accepted liability and advised me that we would not be involved. I reported the matter to the Australian Staff in Washington, DC where there was an immediate over-reaction and I was told to move my family immediately to Canada to avoid any possibility of a court case.

Having previously been declared a pariah by both the Artillery and Armoured Corps much earlier in my career, I was now being hustled out of the USA in a big hurry to avoid embarrassment to Australia.

Luckily we were pretty well packed by this stage so, under arrangements made by Washington, we moved in two limousines, one for us and one for luggage, to Newark Airport and thence to Ottawa where we were met and housed by Major 'Monty' Montgomerie and his wife Phyllis. They were extraordinarily kind and helped us through what was a rather traumatic event. The arrangement was that we would stay in Ottawa for a week and then fly to Vancouver to meet up with the *Orsova*.

A few days in to our stay, we were invited to dinner by a Canadian family (their name escapes me) who were about to leave for the UK to work on MALLARD. After dinner, we were being driven through one of the large parks in Ottawa when we were side-swiped by another car fleeing a tow truck. We all, including the other car driver and his female passenger, suffered cuts and abrasions and were taken to a local hospital for treatment. The offending driver was quite unconcerned about the injuries to my children which incensed me. I said to the Mountie guarding him to keep a close watch or I would do the offending driver a damage. The response I got was 'Don't worry Sir, we will see to it'. I understand that when the driver appeared in Court next day he was sporting bruises from walking into a door. I must admit we were all shaken by the experience but I really felt for our host who was about to depart for the UK and had had a buyer for his car. The next step in the journey was to fly to Vancouver to await the arrival of the *Orsova*. I hired a car and we did some sightseeing and generally revived our spirits. When the ship arrived I drove down to collect the Tinklers to give them a quick tour. On the way a speeding driver went through a Stop sign and ran into me. I was unhurt but the car was a real mess. I was able to get back to the Hire Car firm and announced their car was bent. To my surprise, they were completely unconcerned and did not even retain my deposit.

We had now suffered three mishaps within two weeks and I thought this should be the end of it – more was to follow.

The next day we set sail for Sydney via Honolulu and Fiji. We had two cabins in First Class. The cabins were adequate but somewhat cramped and I hate to think what the conditions were like for the Second Class.

Life on board was very pleasant with lots of activities for the children and entertainment for the adults. We dined in 'black tie' every evening with dancing and competitions to follow. Yvonne and I managed to become the reigning champions of the Limbo dancing competition and won various other prizes. All was going well until Kerry contracted chicken pox which meant she had to go into isolation with a parent in continuous attendance. This put a bit of a dampener on our enjoyment. So we broke the three times bad luck rule and acquired a fourth.

I have very little memory of the visits to Honolulu and Fiji but I do have a clear memory of growing excitement at the prospect of coming home to Australia. America was a great place to live and I had found the work stimulating and enjoyable. I had made many friends, military and civilian, among the Signal Corps of the US, UK and Canada and had established a reputation as a competent engineer and staff officer. A favourable exchange rate and good allowances also meant that we were returning to Australia in a much better financial position than when we left.

I do have a clear memory of rising early on the day of arrival for the first sight of Sydney and shedding a tear when the Bridge came into view. Some hours later we disembarked to be met by my parents and mother-in-law. It was a joyous homecoming after almost two and a half years away.

CANBERRA

I had been posted to the Directorate of Signals as the SIGS 5 A which was the second tier of the technical side working to Lieutenant Colonel Ray Clark under the Director, Colonel John Williamson. The Directorate was a large one with three Lieutenant Colonels, twelve or so Majors plus other staff. We were both a staff directorate and a Head of Corps and there was always plenty to do.

After a short delay, we were allocated a quarter at Cygnet Crescent, Red Hill – a two-storey duplex among a number of government owned properties but a good address! The children joined the St Bede's Primary School, Red Hill, run by the Good Samaritan Sisters (the same Order that had taught me at Glebe). I quickly became involved in the Parish and joined the St Vincent de Paul Conference there.

We had only been in Canberra a few weeks when the SO2 (Officer Postings) told me that he would propose I attend Staff College in 1970. My family were not keen to move again so quickly so I sought and gained deferral to 1971. This decision worked very much to my detriment that I will describe in context later.

I must say that my family and I were enjoying Canberra. Grandparents were reasonably close in Sydney and we made regular visits. I had bought a second hand Ford Falcon on return to Australia which provided a reasonably comfortable ride and had adequate power for easy touring. However, the road to Sydney was still a challenge and took anything up to 5½ hours for the trip.

The Vietnam War looked like going on for a long time and not a great deal of planning went into rotating officers into the theatre. From memory, Captain Dick Twiss was the first of RASigs to be deployed in Vietnam on Signals duties – others had already been posted to the Training Team. Dick was sent up to install the rear link communications and stayed on as Troop Commander. The rear link troop soon became a Squadron and later there was also a Task Force Signals Squadron plus a Chief Signals Officer. In the early days the Postings Officer started working up the seniority list until a few of the seniors complained they would miss out so the sequence was reversed leaving a gap in the middle. I expected to be posted to Vietnam in 1971 but instead was informed that I had been selected to attend Staff College in that year but that I would probably go in 1972.

Working in the Directorate was completely different to SATCOMA. In general terms we had no direct control of our budget and had little to do with actual project management. I did, however, enjoy the work and felt that I was making a contribution to our engineering effort.

I was the only officer in the Army, and, in fact, in the whole of the Defence Force, that had had direct experience with satellite communications and I was called on from time to time to provide briefings. I attended the odd seminar/conference on satellite communications and soon discovered that military systems were far more advanced than civilian ones. I was often unable to contribute to open forum discussions as much of what I had learnt was security classified.

Military systems were also more advanced than civilian ones in other areas of communications including message and circuit switching technology, radio transmission and spectrum engineering. Unlike today, cryptography was unknown outside government agencies. Exposure to advanced systems made military officers very attractive to industry and RASigs lost a lot of very competent officers due to this fact.

Apart from being the resident expert in satellite communications, I was mainly involved in development of the Australian Communications

Network (AUSTCAN) which consisted of a High Frequency Radio Network with message switching by STRAD. I was also involved with Combat Net Radio (CNR) in the HF and VHF bands. This meant that I had a lot of dealings with contractors, mostly over long lunches!

My main contacts were with Racal and Plessey. Racal were major suppliers of CNR but they were also very interested in bidding for transportable high power HF radios for what was known as the AUSTCAN extension project. Plessey were interested in developing a presence in mobile area networks as the writing was already on the wall for the demise of MALLARD. I developed a good working relationship with the representatives of both these firms which was of benefit later in my career.

During my time with the Directorate there was a deal of manoeuvring within Racal staff for their top job in Australia. The final winner was Bruce Goddard who had been one of the lecturers when I was studying for my postgraduate degree. Bruce and I became good friends and he started courting me as a prospective employee. He made me a good offer but I remained committed to an Army career.

Early in my posting at the Directorate, I found that little was known about facsimile as a technology even though digital facsimile was intended to be a major component of MALLARD. I did a quick search of suppliers and found a Sydney based firm that distributed an advanced Japanese model (the actual brand escapes me). I arranged for the dealer to come to Canberra to demonstrate his equipment. A great deal of interest was aroused and the dealer was ecstatic. A week or so after the visit, the dealer called me and asked if we could meet on my next visit to Sydney. I readily agreed and he invited me to lunch. As the lunch progressed, he said that I had achieved more for his business than any of his sales staff and he would like to give me a present. I said that this was unnecessary as new technology was part of my job. I half expected a pair of cufflinks or a pen with a company logo but, to my great surprise he said that he thought a briefcase with $5000, in small notes of course, would be suitable. I had never been offered a bribe before and rejected the offer out of hand. I immediately wondered if this was some sort of

ASIO sponsored test and reported the whole matter on my return to Canberra. I never heard any more on the matter until a year or so later, two security officers visited unannounced at Staff College to interview me but left when they found we were all out on an exercise!

Towards the end of 1970, I thought my career would come to an abrupt end. I had been preparing papers for the purchase of 10 high power transmitters for the AUSTCAN network and had sent a Minute to Defence, which I signed for the Director, seeking $M10 for the purchase. A day after the Minute left my office, I discovered that I had made a substantial error – I should have sought $M1 NOT $M10. As an aside, in those days budgets and estimates were still being shown in $K so the $M9 error was substantial. In some fear and trepidation I went to the Director and confessed my error. He was completely unfazed and told me to send another Minute referring to my erroneous one simply saying delete 10 insert 1. He then went on to tell me that most people did not understand millions of dollars but readily reacted to errors in smaller sums. Many times in the future I was to find this to be quite true.

STAFF COLLEGE QUEENSCLIFF

Towards the end of 1970, my posting to Staff College was confirmed so we prepared ourselves for another move. There was limited housing for families at Queenscliff and much effort was expended by staff to facilitate students acquiring suitable rented accommodation. The Army assisted by funding students to visit Queenscliff in advance of the start of the course to select a house and settle leasing arrangements. I was able to obtain a lease on a nice house in Port Lonsdale which was within a few kilometres from the Staff College. Ownership of the house had changed just before the lease was signed and we were assured of occupation until the end of the year. The new owners subsequently changed their minds but more of this later.

Qualifying at Staff College was not a guarantee of promotion but it certainly helped! Staff College was intended to prepare officers for Grade 2 (with the rank of Major) staff appointments but most of us attending were towards the top end of seniority as Majors and were approaching the time to face the Promotion and Selection Committee for promotion to Lieutenant Colonel.

Attendance at Staff college was much prized, not only because of the enhancement of career prospects but because it was very much a sabbatical year allowing reflection and the opportunity to examine issues of interest. This was certainly the view of the Commandant, Brigadier Bogle, but many of the Directing Staff were determined to fill up every spare period with formal instruction or discussion.

The Directing Staff (DS), some fourteen in number, were all Lieutenant Colonels who were well qualified and highly respected within the Army system. Among the 14 were one New Zealander, one Canadian, one American and two Brits. There were also two staff officers, an adjutant, a quartermaster and a Mess Secretary.

There were seventy in the student body including 17 from overseas. We were a mixed bag covering all Corps and were mostly contemporaries in age and seniority. The overseas students came from New Zealand, UK, USA, Canada, Thailand, Brunei, Ceylon, India, Pakistan, Indonesia and Malaysia. We also had a Public Servant and a RAN officer. We all recognised the importance of the year and we were all prepared to work hard. We also knew from former students that we would all play hard as well.

Having been virtually out of the system for two and a half years while in the USA and then being pretty much restricted to Canberra while on the staff, I had lost touch with many of my contemporaries and so was looking forward to renewing old acquaintances and making new friends. Of the seventy students, I knew twenty quite well but I had never met any of the staff.

It was also to be a great social year with a nearly constant round of parties, dinners, etc in addition to the rather full Mess programme. Entertaining became very competitive among some of the students and became quite a drain on living expenses. I also understand it was not unusual for some flirtations to go on but I must say I didn't observe any. There were ample opportunities for sport with golf, hockey, tennis, squash and badminton being well patronised. The ladies were not left out and they had programmes covering the same sports plus Fencing and also Mahjong, Bridge, Art and Charity group. There was also a mixed Drama Group.

As 1971 started, the anti Vietnam War protests were becoming more and more strident and even the Liberal parliamentarians seemed to be tiring of the war. However, it was the only war we had and much of the Course was conducted against the background of Vietnam.

It was on our first working day at Queenscliff when the whole student body was assembled in the main lecture hall (a long narrow building with a non-sloping floor) to be addressed by Lieutenant Colonel Robin Hone. Having spent much time in teaching establishments, I immediately scampered to the back row. Robin's actual appointment was General Staff Officer Class 1 but he was known as Officer Commanding Ladies Happiness because he looked after, inter alia, the spouse programme.

Robin gave us a long lecture about the programme, what was expected of us, the idiosyncrasies of the Commandant, Brigadier Bruce Bogle and much other essential information. At the end of his talk he asked the assembly if we had any questions. I put up my hand whereupon he yelled at me:

"Stand up when you ask a question."

I immediately answered:

"I am standing up, Sir."

Which reduced the class to peals of laughter and set up a sequence that was followed, not only at the College, but whenever I rose to address an Army audience for the rest of my career. I played this to the hilt and soon found that I could reduce my presentation time by a minute or two by going through the pantomime. The pantomime did not finish here but followed me throughout my Army career and into retirement.

Instruction was based on the well proven method of central lectures and syndicate discussions. We were formed into syndicates of ten under a member of the Directing Staff. Syndicates were changed each term so we each had close contact with four different members of staff.

A great deal of emphasis was laid on improving our English expression so we spent many hours writing papers on all sorts of subjects in addition to honing skills on producing such military documents as Appreciations and Orders of different kinds. All our written work was by hand and the use of ball pens was discouraged as such implements did not allow one to express mood (heavy downstrokes, etc). In line with this emphasis on expression, our two major reference books were the Concise English Dictionary and Fowler's Modern English Usage.

We also spent a lot of time on Tactical Exercises Without Troops (TEWTS). This involved being bused into the field at various locations in the region where great battles were fought without loss of blood or reputation. Such exercises required considerable skill at map-reading which was a particular fetish of Lieutenant Colonel Tom (I am not a black hat) Williams, 14/20th Hussars who was very quick to send us scurrying up any nearby high ground to get our maps properly orientated.

I must admit I was not very good at Tactics, particularly when required to draw conclusions from a study of 'the ground', but I did learn that 'Two up, up the guts with smoke and tanks' was never very far from wrong. For the uninitiated, the above sentence translates to two companies forward making a frontal attack under cover of smoke screens provided by the Artillery and support from Tanks.

One of the great features of TEWTS was lunch in the field. At the appointed time we would be transported to a suitable location where the Mess staff had set up trestle tables with silver, etc so that we could eat in style.

From time to time we would be subjected to Visiting Lecturers. These varied from the brilliant through all degrees to absolutely awful. Most of these lectures are now but a dim memory but one stands out. We were 'honoured' by a visit by a Major General who had recently returned from commanding Australian forces in Vietnam who was to present a lecture on Joint Command. The lecture was classified so all foreign students were sent off elsewhere so that only we ANZACs could have the benefit of great wisdom. After the niceties of the student body standing to attention when the great man entered and the introduction by the Commandant, we took our seats, the lecture hall lighting was dimmed and after some opening remarks along the lines that as the Joint Force Commander he had given considerable thought to how a joint force should be commanded. The call was then made to the slide operator:

'Sergeant, Slide One.'

We waited with bated breath to see:

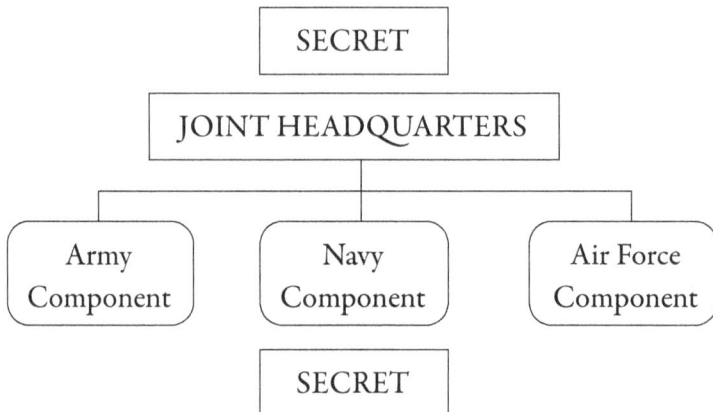

```
                    ┌─────────────────┐
                    │     SECRET      │
                    └─────────────────┘
            ┌─────────────────────────────────┐
            │      JOINT HEADQUARTERS         │
            └─────────────────────────────────┘
   ┌──────────────┐   ┌──────────────┐   ┌──────────────┐
   │    Army      │   │    Navy      │   │  Air Force   │
   │  Component   │   │  Component   │   │  Component   │
   └──────────────┘   └──────────────┘   └──────────────┘
                    ┌─────────────────┐
                    │     SECRET      │
                    └─────────────────┘
```

This was the one and only Slide and the student body was extraordinarily polite with no laughter and absolutely straight faces when obligatory questions were asked. I must say that we were all extremely pleased that this vital knowledge did not fall into the hands of the foreign students.

Early in the course, I decided to try my hand at golf and had the good fortune to receive some coaching from the RASigs DS Lieutenant Colonel Peter Carruthers. Peter was a great sportsman and he was surrounded by a considerable amount of Signals folk law. The best story was of his appearance before a Promotion and Selection Committee for Lieutenant Colonel. The interview went very well and promotion seemed in the bag. At the conclusion of the interview he was asked the obligatory question 'Do you have any problems?' Peter jumped to his feet, took up the appropriate stance and replied 'Yes, I'm having difficulty with a recurring slice'. The Committee was not amused and regarded the question as flippant, there not being a golfer among them, and his promotion was delayed for some years!

The course was going well for me and I found the work interesting and challenging. My written work invariably attracted good marks and I was getting the hang of tactical problems and such skills as map marking, artillery fire planing, Appreciations, etc and such intricacies as calculating 'time past a point' of convoys. My public speaking and presentation skills were improving and the 'I am standing up' routine had become well honed. All in all, life was pretty good when I had a recurrence of an old medical problem which caused me a great amount of discomfort and caused my student performance to go down hill rapidly.

After some weeks, the visiting medical officer decreed that I needed surgery and I was admitted to a hospital in Geelong run by Catholic nuns. On the day of admission, I was rather dejectedly sitting by my bed anticipating the surgery to follow the next day and the discomfort that was to follow. A fellow patient said something along the lines of "You look a bit down, would you care for a drink?". When I replied that I certainly would, my companion rung the nurses bell and when the Sister arrived asked for some ice. This was supplied and I was presented with a very welcome Scotch. It transpired that the nuns considered that, as it was highly unlikely that anyone was going to die in the 'bum ward', alcohol was preferable to painkillers and sleeping pills! Following a week in hospital I was able to get back on track although my sporting activities, golf, hockey and badminton, were somewhat restricted.

We played badminton in a hall at the bottom end of town and invariably followed this with a few beers at a pub frequented by the local fishermen. The badminton players were a mixed group but it included a few of us who were well aware of the existence of a highly classified establishment that existed at Swan Island, just across the causeway from Queenscliff. Those of us in the 'know' never discussed the place being well aware of the provisions of the Official Secrets Act and were shocked to overhear one of the fishermen announce in the pub:

"Spy school is operating again – there was a submarine hanging around last night."

So much for security!

One of the high points of the year was the battlefield tour of PNG. We were loaded into a Hercules and headed north to Port Moresby. In Port Moresby we were quartered, because our number exceeded the capacity of the 1 PIR Officers Mess, in the soldiers barracks which had been scrubbed and polished and were more than acceptable. We had the usual round of briefings and presentations including a comprehensive talk on the rivalry that existed between the Army and the Police. This rivalry was seen as most beneficial to the stability of the country. As an aside, both the PIR and the Police were well disciplined, well turned out and good at their respective jobs.

Among other visits on the programme, a sombre highlight was our visit to the Bomana War Cemetery. We saw at first hand the great tragedy of the New Guinea campaign with many members of the same family buried side-by-side.

The Police were also on the programme and we were treated to a display by the anti-riot dog squad of the PNG Police which was truly amazing. On the handler's command, his dog would race to the designated target, knock the legs from under him and then lay on his chest growling but without making teeth contact.

After a few days in Port Moresby, we flew to Mount Hagen where we attended a sing-sing and were royally entertained by the local expatriates. We continued our tour of PNG in a pair of Caribou transports flying low with the back ramp down so we could easily see the WWII airfields, battle sites, etc.

I'm unsure of the where in the tour sequence it occurred, but we landed at Wau airfield for a briefing on one of the major battles. Landing there was and still is exciting indeed. The airfield is on a steep slope and the plane is landed uphill. The pilot must take extreme care and make a right angle turn to park the aircraft before it rolls downhill! The battlefield briefing was supposed to be given by one of the Australian Army staff from Port Moresby. I gather the officer in question became heartily sick of presenting the same material time after time, so he prepared an audiotape which was meticulously researched and well delivered. We students were duly lined up on the landing strip and oriented to the mountains for the talk to begin. Unfortunately, the mountain was shrouded in cloud so the directions to look to the front, left, right, etc were of little value. Much mumbling went on among we students and the presentation was finally aborted with the accompanying DS making veiled threats about wrath descending from above on 'bolshie' students.

We also visited 2PIR in Lae. In stark contrast to the trouble taken by 1PIR, we were housed in a less than clean soldiers barracks with no great effort made for our comfort. We were, however, given a good dinner with lots of drinks to follow before trundling off for sleep. During the night one of the students was 'groped' by a native soldier. The alarm was raised with much shouting and running about but the culprit escaped undetected. During the hullabaloo, it was discovered that my bed was empty and grave fears were held for my safety. I was soon discovered in what should have been a vacant bed dead to the world and oblivious to the noise. I had in fact got up to answer a call of nature and had become disoriented on my way back and climbed into first empty bed I found. This event led to the Best Remembered For entry in Tookarook as: 'For his tropical disappearing act'.

After a great trip we flew back to RAAF Laverton in our Hercules. Towards the end of the journey, I was on the flight deck on finals in very bad weather when the pilot threw up his hands and yelled out 'I hate landing these things' – no doubt to scare me. He certainly succeeded but then the boot was on the other foot when a windscreen wiper flew off which certainly got the pilot's undivided attention!

About midyear I was informed by the Director of Signals that I was to be posted to SVN to take over from Peter Wilkins as Squadron Commander of the 110 Signal Squadron which provided fixed in-theatre and rear link communications. I was expecting this and quite excited at the prospect of operational service. In anticipation of this,

I purchased a new car, a Ford Fairmont, so that my wife would not have to worry about car reliability while I was away. More on this later.

We were not long back in Queenscliff when the owners of our rental house announced they wished to occupy and served me with an Eviction Notice. This caused me no end of concern as it would be a major disruption in a critical part of the course. I informed the staff and Peter Carruthers took up the cudgel and appeared in court for me. He eloquently put my case and the judge decreed the eviction order would stand to be executed two weeks after the course finished.

Late in 1971, war broke out between India and Pakistan. This caused both staff and students much anxiety as we had students from the protagonist countries on the course. The contrast between them could not have been more stark – the Pakistani was a meek and mild Service Corps officer while the Indian was a Sikh in the Armoured Corps. To their credit, both officers maintained polite relations but the Sikh was all for packing up and going home to fight. I'm not sure how things were managed, but a senior official from the Indian High Commission came to Queenscliff to order our Sikh to stay put.

Talking to the Indian about the event was quite enlightening. He assured us that in the previous wars between India and Pakistan, all and sundry left headquarters and support areas to head towards the fighting. He told us of general officers who left their posts to go to the front perfectly willing to act in junior officer positions in their old regiments. He saw nothing wrong in denuding headquarters, etc to go to the action. He declared that no one heading towards the action would be stopped but anyone coming back from the front would be summarily shot!

I had the pleasure of being teamed with our Indian to play the enemy in our final telephone battle, which involved a river crossing, to wind up the course. He was very experienced in matters tactical but, unfortunately, he found it impossible to accept that our job was to move the 'enemy' resources in such a way as to exercise the others rather than fight to win. When I said we could not make a pre-emptive crossing to catch the 'Blue' forces and destroy them in place, he refused to make any further contribution and left me to it.

The anti-war protest movement was gaining ground as it was in the USA and Prime Minister McMahon was moving to extract Australia. I still had my Posting Order and wrote to the Director of Signals, Colonel (later Major General) John Williamson asking that I be given a week to settle my family in Canberra and then go directly to SVN. He advised me that I would have to first go to Canungra for pre-deployment training but that he would do his best for me. Not long after this, the PM announced that there would be no more reinforcements sent to SVN and that those in theatre would be extended until the complete withdrawal of forces. Soon after this I received a revised Posting Order appointing me as a Grade 2 Staff Officer in Personnel Branch at Army Headquarters. So my decision to defer Staff college in 1970 now meant that I would miss my opportunity for operational service with the only other prospect being as an United Nations observer.

The last social function for the year was a dinner dance which was attended by all students and included most wives who had not moved to Queenscliff for the course. Included in this number was the wife of Bert Irwin who was a very tall statuesque well endowed woman of pleasant personality. In the course of the dancing the Commandant, Brigadier Bogle, showed a rare show of humour when he suggested I should dance with Bert's wife but added 'If you move your head you are on a charge'. I accepted the challenge and ended up with my head firmly planted under her armpit and so unable to move.

On our day of departure we went to our boxes for the last time to collect our Staff College Club ties and receive our final report. Peter Carruthers had predicted I would receive a 'B' Pass but, alas, he was wrong so I had to settle for the mediocre 'C'.

I very much enjoyed the year and was now looking forward to taking up my new appointment in Canberra. It was and still is considered to be an essential career move to have a non-Corps staff posting so I was pleased to have been given a posting as a Deputy Assistant Adjutant General (a much more impressive job title that Staff Officer Grade 2 Personnel) in the Officer Policy section of the Directorate of Personnel

Plans. My mother was most impressed with the title and told every one in the street that Peter had been promoted to General. I was more than happy to be going to Personnel Branch as with my technical background I had half expected to be posted to Material Branch.

BACK TO CANBERRA

On our return to Canberra we were allocated a government house in Curtin. This was a two-story duplex very similar to the house we had occupied in Red Hill before going to Staff College. I expected that we would have at least a few years in location so I quickly got involved with the local Saint Vincent de Paul Society and Holy Trinity Primary School which both Damian and Kerry attended. As the year progressed, Damian and Kerry took up Scouts and Guides respectively and both joined Hockey clubs. As a result, I became the Assistant Scout Master and a coach of a girls hockey team.

Kerry had developed asthma which was a great worry as Yvonne's brother had died as a result of the disease in his early twenties. Yvonne came across a South Coast doctor who promoted a revolutionary approach to the problem and was instrumental in forming an Asthma Society in Canberra in which she was heavily involved throughout our time in Canberra.

There were a number of my RMC Classmates living in Canberra as well as colleagues from the Corps of Signals, so we had an established circle of friends and enjoyed an active social life.

But now to work – my immediate boss was Lieutenant Colonel Tom 'Hotto' Flannagan so nicknamed by his RMC Class (1954) for his penchant for borrowing anything that wasn't nailed down. 'Hotto' was not the sharpest officer I knew but he was reasonably conscientious and had the good sense not to interfere with the work of his subordinates. 'Hotto' had married well and commuted to Sydney each weekend. Not long after I arrived, he and his wife had purchased a string of terraced houses in Paddington which they planned to renovate. When 'Hotto' told me he planned to personally renew the wiring of the houses, I was horrified and proceeded to tell him of the dangers of the work and difficulties he would have with insurers if problems occurred. This was my first foray into engineering consultancy which was, of course, unpaid.

Our Director, Colonel (later Brigadier) John Salmon had recently returned from Vietnam where he had planned the withdrawal. He did this task extremely well and was rewarded with a CBE. I had know John since RMC days were he had been, inter alia, the instructor in Staff Duties. He certainly practised what he had preached and was an absolute stickler for crossing 'T's and dotting 'I's. Any Minute for his signature had to be absolutely right and he could pick out a misplaced comma or incorrect spelling at a hundred yards. Erroneous Minutes were returned to the Typing Pool for correction which often introduced other errors with further returns. The record for this was held by Peter Wilkins who produced 47 drafts of a one and a half page Minute before it was signed.

National Service was still in operation but being wound back. We had two conscripts in the Directorate and they were like chalk and cheese. Both were university graduates. One was a very disgruntled clerk who took great delight in sailing as close to the wind as possible. He was not a happy soldier and we were all pleased to see the last of him. The other was a Lieutenant who held a PhD in mathematics and who went about with a permanent wide grin on his face. I soon learnt that he had only recently returned from Bangkok where he had spent the best part of a year living in the Palace and coaching the Crown Prince in mathematics before the Prince started his course at RMC. I cannot imagine that anyone else could match this for an enjoyable stint as a conscript!

I quickly became very interested in the field of policy development and certainly found it more interesting than the other important but rather mechanical aspects of the personnel function. I enjoyed the intellectual rigour required of policy development and found the demands of providing concise briefs and quick, well written and presented responses demanding but enjoyable. I seemed to have a modicum of talent for the work and soon became the unofficial personal staff officer to the Director as well as doing my own work.

One of my many responsibilities was the allocation to Corps of officers graduating from RMC, OCS and OTU. On the face of it, this was a mechanical problem based on percentage representation of the various Corps within the Army and this is how it had been treated in the past. This approach had led to many anomalous situations particularly in the availability of officers of some Corps to fill non-Corps postings. I developed a very simple mathematical model which simplified the allocation process and went a long way towards ironing out most of the anomalies. The model greatly impressed my superiors, most of whom were not greatly numerate, and enhanced my reputation.

As part of my exposure to the problem of allocation to Corps, I was detached from the Headquarters to be an observer on the selection interviews for the Officer Cadet School. I took no part in the actual process but sat behind the Selection Board watching and listening. The Board I sat in on was chaired by the Commandant Colonel (later Major General) David Butler. Prospective cadets were brought in one by one and asked various questions about their sport, academic achievements (these were not greatly valued), back to sport, their ambitions and then rounded off with other questions on their sporting prowess. It soon became obvious that the major factor in selection was the position the contender could fill in the OCS Rugby Union team. Excellence in Australian Rules Football was grudgingly acceptable but did not compete well with comparatively lower talent in Union. At the time I thought that this was not a great way to select future leaders but on reflection and after watching how cadets selected on this basis performed at OCS and later in the Army, I changed my view!

Another area that I became involved in was the vexed question of return of service for education and training that had a marketable value. One of the issues driving this was the high wastage rate of officers to industry. This included general service officers and not just the engineers, project managers, pilots and other specialists. The dilemma was that the ordinary training and education that officers were given were also essential for the efficient operation of the Army. For example, staff training at the Staff College was essential to the Army but also had value for industry. There was a wider view that skills that were transferred to industry were also in the national interest but care was needed as a too stringent return of service obligation was a disincentive for officers to undertake training essential to the Army. After much to-ing and fro-ing, I finally received approval for my proposed 'One for One plus One' policy which meant that a one year course with commercial value attracted a two year return of service obligation. This policy was still in force in the late 80s and may still well be.

Another difficult policy area for the Army and hence for Personnel Branch was what to do with members of the Womens Royal Australian Army Corps (WRAAC) who managed to become pregnant. In earlier days, immediate discharge was the solution but this was becoming increasingly unacceptable in the general community and so provisions had to be made for pregnant soldiers. Suitable uniforms were designed and policies on leave, etc were developed. There was a great need for women in the Army, particularly in Signals and Transport, and much effort was put into recruitment and retention.

During my time in the Branch, the Director WRAAC was Colonel Kath Fowler. She was a thorough lady and absolute innocent. She was much admired in Signals for her outstanding war time service as an intercept operator. She was also quite unworldly and often during personnel meetings came out with statements with very much unintended double entendres causing most of us sitting around the table to have great difficulty keeping a straight face!

Colonel Salmon, previously mentioned for his meticulous staff duties, was also a man who demanded a serious work ethic. He expected staff to arrive early and stay late and considered we should take a minimum break at lunch time. Going out for lunch was frowned on and we were expected to seek his approval for such frivolities. When a request was made the usual less than enthusiastic response was 'Well, if you really MUST go' which was intended to make the supplicant feel really bad. I have no idea how this penchant came to the attention of the Adjutant General, Major General Tim Vincent, not known for his sense of humour, but it did and this led to the following incident that greatly amused the Personnel Plans workers.

It was a Friday afternoon and the Director had arranged for his wife to collect him outside Russell at 16:50 before they drove to Sydney for an important cricket match. At about 16:45 a call came from the AG for our Director to present himself in the General's office. As usual I was taken along with notebook and pencil to sit behind the Director and take any necessary notes. We were ushered in to the presence and were invited to sit. The General looked down to papers on his desk and continued writing. After a few minutes, the General looked up and said *"What are your electricity bills like John?, mine seem to be getting bigger and bigger"*. He did not wait for an answer and continued writing. Eventually, it got too much for the Director who said *"Sir, I have my wife waiting for me outside and I need to get away to Sydney"*. The General then looked up said *"Well if you really MUST go, John, you go"*.

The return walk to the Director's office was conducted in fuming silence while I struggled valiantly to retain a straight face. I regret to admit that I was not very loyal to my Director and related the whole incident to my brother junior staff officers who all but collapsed in paroxysms of laughter.

My Officer Policy appointment also involved me in preparing submissions to the Minister on many and varied issues. This was a very time consuming business as the actual submissions needed to be on a single page with any supporting documents/references included as annexes. All papers had to be suitably tagged and neatly placed in folders. Responses often had to be agreed across various responsible agencies and so it was often difficult to meet the time objectives set by the Minister's office. In performing such tasks, I was extremely grateful for the training I had received at the hands of Lieutenant Colonel DAC Griffith when I was a young Captain on his staff in Sydney in the 60s.

Apart from 'Ministerials', there was often a need to brief the Adjutant General on various submissions made to him by outside agencies. One of the most memorable that I became involved with was a letter from the President of the Imperial Services Club requesting that all National Services officers be promoted to Captain before they were discharged. The President was greatly concerned that it would be harmful to the prestige of the Club to have on its Membership Register, ageing Second Lieutenants! Unfortunately, I never did see the response but the rejection of the request was couched in such language that no further correspondence was received.

It was not all hard work at Russell although we did work unconscionable hours from time to time. Sometime in mid 1972 an Army Office Wine and Food Club was formed. We took it in turns to produce a meal that was consumed in the tiny Mess on the ground floor of 'G' Block along with wines of interest. I'm afraid that, at times, more wine was consumed than what was required to comment on their various characteristics and merits. We were generally very discreet but I remember one occasion when the Chief of the General Staff sent his ADC to tell us to clear out before he came down personally. Unfortunately, we had not realised that the air conditioning duct in the Mess had an outlet in the great man's office two floors above!

During this time I became very friendly with Graeme Burgess, initially through our association with Scouts and later within Personnel Branch. Graeme and I developed a liking for poker machines and often crossed the border for some harmless fun.

During the year I joined the Knights of the Southern Cross and attended many functions with Tony Devoy who became a close friend. I also renewed friendships with Neil Harris and other members of Signals. Neil had come up through the ranks to the rank of Major and was the resident expert on Pay and Conditions of Service. He was also a dab hand at cards and most lunch times with joined with a few others to play a version of 500 called 'Black Fellah's Bridge' – very non-PC I'm afraid.

Some time in late September early October I received advice that I was to appear before the Promotion and Selection Committee for Lieutenant Colonel on 26th of October. This was the first real break point in an Army career as prior promotions had been basically automatic. The Committee had access to all previous Confidential Reports and the recommendation of the Head of Corps. My reporting history was good and I had the advantage of a good technical education and had had a good pass from Staff College. However, there were limited vacancies to be filled with an over supply of candidates. I had had very limited time in command and no operational experience, so I was a little nervous about my prospects.

All candidates knew we would be quizzed about our reports, asked the usual 'what would you fix' question and be examined on our knowledge of world events. So we all went to work on reading Newspapers, Time and the like with much effort in trying to second guess the Committee. I had been told many times that the 14 years of reports would pretty well decide my fate and that the interview by the Committee was simply to allow them to confirm their decision. This all seemed pretty logical, but, like most, I spent many sleepless nights wondering if my career was about to grind to a halt.

Finally, the day came and I was ushered in to the interview room. I have no recollection of what I was asked or how I answered but I do remember that I had received no hint of my fate. So it was now a time of waiting and with no real indication of how long the wait would be.

The year ground on and after a visit to family in Sydney for Christmas it was back to work in Personnel Branch and to see what 1973 would bring.

In March 1973 I was detached to Defence for three months to work on a study of satellite communications for the ADF. There were three of us, Lieutenant Commander Rick Patten, a Squadron Leader whose name escapes me and myself. As I was the only one who had had formal training in the field, I chaired the group. The task was straight forward enough and we had much too much time for the task. Rick and I enjoyed the detachment very much but I cannot say the same for our RAAF colleague who was a bit of a stickler and did not care for long lunches.

Having worked at the US Satellite Communications Agency in

the early days I had experienced at first hand the great advantages of this form of communications but I was also aware of the dangers in total reliance on this mode. So in preparing the study report I was at pains to promote the idea that research and development into high frequency radio should be continued. I also expressed reservations on sharing transponder space on US satellites notwithstanding the fact that the ADF would have difficulty justifying a Defence satellite at that time. I do remember raising the possibility of a civilian/defence shared satellite but the requirement to handle different frequency bands would have stretched technology of that time.

We produced the report on time and it was well received by the Director General Joint Communications. Nothing much was done with it and I will take up the story again when I joined Joint Communications in 1979.

In those days promotions could only occur into a vacancy so those of us in the pipeline would look carefully at who was retiring and who was being promoted from Lieutenant Colonel to Colonel. I'm not sure who was the first in my RMC Class to be promoted but it certainly was not me. I think Geoffrey Christopherson may have been the first in the Class promoted with Mike Jeffery following soon after. These promotions were early in the year (1973) and a number of other announcements soon followed. I was starting to worry when I received a call from the Military Secretary to say that I was to be promoted on the 2nd of April and would be the Staff Officer Personnel in the Directorate of Signals as soon as the study was completed. The Director was Colonel (later Brigadier) Frank Burnard.

The Directorate of Signals was unique in the Army as it was both a staff directorate providing advice to the General Staff as well as being a Head of Corps. It also had responsibilities for cryptographic equipment and material as well as overall technical direction of the Army Communications Network. I'm not sure of the timetable for this but the Directorate split with the Director staying in Russell Offices and my Personnel Section relocating to Campbell Park Offices. At the Director's insistence, my title became SO1 Head of Corps which did not please many of my contemporaries in the Corps. At the same time a new organisation was formed, the Army Communications Agency, under Colonel (later Major General) Barry Hockney, which took over the entire responsibility of the fixed communications network.

I found this posting particularly demanding as every decision taken on an individual could have a marked affect on their careers and on their families. The approach I took for both officers, NCOs and soldiers was to involve the individuals as much as was possible in determining their imminent posting and their future aspirations. This invariably worked although there were a number of occasions when individuals ignored my 'at this stage my intention is' rider and entered into arrangements that caused them loss of face and often financial loss. As one can easily imagine, any one in my or similar positions, attracted both bouquets and brickbats.

I think towards the end of 1973 but it could have been later, the Directorate took over the career management of those WRAAC who worked with RASigs. We even went to the extent of re-badging both officers and other ranks although some officers exercised their option of remaining with WRAAC. I consider that this was a major achievement and it did much to enhance the promotion prospects of female soldiers and officers.

I had a very competent staff who, inter-alia, developed an individual record system that was adopted by many other Heads of Corps and eventually became the standard system across the Army. From the best of my memory, the folder system was mainly the work of Warrant Officer John Brown. With their help, I had a very efficient and well regarded section which did a lot for my future career.

At the end of September, I packed my bags and went to the Jungle Training School, Canungra, to undertake my TAC5 Course. This was a necessary course required under the Defence Act for substantive promotion to Lieutenant Colonel. By the time I got to do the course, it had ceased to be an assessed course which was a great relief as I was not overly good at tactics. In previous years the TAC5 had been the undoing of many fine officers as failure to qualify after three attempts precluded promotion. TAC5 could be brutal and I regret to say that some of the Canungra staff revelled in the power and were quite cavalier with assessments. In many circles it was believed that the main requirement to pass TAC5 was to have failed once before.

With the pressure of assessment gone, the course members settled down to enjoy the break and to learn from the experiences of fellow course members. It could be quite amusing at times when some staff members lacked the experience of the students and heated arguments ensured. I clearly remember an incident when a rather officious and opinionated staff member berated a student on his solution to a particular problem saying 'You obviously know nothing of tactics' which created something of an uproar as the student being berated wore the ribbon of the Military Cross!

During the course we were subjected to a visit by the GOC Training Command, Major General BA McDonald. Just prior to the arrival of the great man, the Chief Instructor collected us students together and pleaded 'please don't hang about in the bar too much and look worried – the General still thinks this is an assessed course'. This we did and all went well.

I had become more and more involved with scouts and was the Troop Leader of 1st Curtin and as such led the troop to the Jamboree in the Adelaide Hills leaving Canberra on Boxing Day 1973 and returning on 8 January 1974. It was quite an event as we travelled by special train via Melbourne – a harrowing journey. I had not been to a Jamboree since 1953 so I found the whole affair a great experience.

Although my appointment was very much oriented to personnel matters, I did manage to keep my hand in on technical issues by attending IREE lectures and reading the technical literature. As part of my desire to remain technically current, I managed to get approval to attend a three day satellite communications symposium at the Academy of Science. The symposium clearly demonstrated to me that civilian knowledge of satellite technology had not advanced much since I had returned from the USA and was still way behind that in the military. I had to bite my tongue on many occasions as some of the more ridiculous statements made by some presenters could only be refuted by reference to classified information. I had a few approaches during the symposium with job offers but I had no interest at all in leaving the military.

A feature of every Director's tour of duty was his Corps Conference. This was invariably held at the School of Signals in Watsonia and involved all unit and sub-unit commanders plus as many of the remainder of Corps Majors and Lieutenant Colonels as it was possible to assemble. The usual aim of the conference was to outline the Director's view of the Corps at the time and to outline the way ahead.

On the opening day of the conference, we waited in anticipation for what the Director would say in his Opening Address. As he entered the room and walked to the dais, from the loudspeaker came the pop tune 'People Who Need People'. Most of us thought the host, CO/CI of the School, Lieutenant Colonel Bill Mostyn, had lost it and we waited for the blast from the Director. Instead he continued smiling and waited for the end of the tune and then spoke to us of the importance of our people. It was really quite refreshing as previous conferences tended to be centred on equipment and tactics.

Towards the end of the conference, the Director joined a group of us heading into St Kilda to sample the nightlife. As we approached St Kilda we saw a huge sign on a tall building housing one of the less salubrious night spots which read 'This Is The Show'. The capitals were huge with the remaining letters small so from afar the sign read 'TITS'. This looked promising so we entered therein. The show consisted of a number of striptease acts and the format was well known to the locals. With the connivance of the Director, I was manoeuvred into the front line of our group. At the start of the third act, a well endowed, scantily dressed performer appeared out of the dark, removed my glasses and inserted them into the top of her G-String. The compère then invited me to retrieve my glasses without using my hands, i.e. by my teeth. This I duly did with howls of encouragement from my colleagues. This somewhat unedifying event entered into the folklore of the Corps and received a mention when I was farewelled from the Corps many years later.

To digress for a moment, Corps identity was very important in the Army at the time and much emphasis was placed on officers supporting Corps Funds that paid for Corps paintings, histories, etc. Subscriptions were collected once per year, usually around Confidential Report time. Non-payment was frowned upon to say the least and generally resulted in adverse comments by senior reporting officers. This attitude was not unique to the Corps of Signals but was in general vogue throughout the Army. I mention this as background to an event that I would have preferred to have missed.

One day I was accompanying the Director through the G Block quadrangle, when we met my old friend Colonel David McMillen. Without any preliminaries the Director launched an attack on McMillen for not contributing to Corps Funds. A heated argument ensured and I attempted to leave the scene but Burnard insisted I stayed put. I found the whole affair very awkward and unbecoming of both of them.

Colonel Burnard was promoted in June and replaced by Colonel Barry Hockney who had held the ACA appointment. I must say that I was sorry to see Burnard go as we had worked well together and I did enjoy his company. Colonel Hockney was Director for just over a year when he was replaced by Colonel RA Clarke in April 1975.

Nothing much changed for me and I was kept busy with running the personnel side of the Corps. It never became less demanding and was always rewarding. The short distance between Russell and Campbell Park Offices might as well have been hundreds of miles and the Director rarely visited so I was left to get on with it. 1975 was the 50th Jubilee Year for the Corps and I became very much involved in planing for it.

My father had been in poor health for some time and had been hospitalised on many occasions for heart disease. I had been at home in Sydney when Dad had had heart attacks so was not overly surprised to be called to Sydney in late November 1974. I saw Dad who seemed to be improving slightly and there was a prospect that, once again, he would recover. I returned home to receive a call from my Mother early in the morning of the 6th December 1974 telling me that Dad had died.

Although Dad's death was not unexpected, I found I was profoundly saddened and had difficulty coping with the arrangements and a grieving mother. Dad was cremated and his ashes placed under a Rose Bush in the Northern Suburbs Crematorium. There was no memorial so I donated a statue of my Patron Saint, St Anthony of Padua, in the Chapel of my old school, Holy Cross College.

Early in the year I had the good fortune to be selected to attend the one and only Canberra based Industrial Mobilisation Course. The course had been instituted after WWII in an attempt to overcome the problems that had arisen during the war that were perceived to have arisen by a lack of mutual understanding between the military, public service and industry. We had a week of orientation at the Naval Air Station, Nowra, followed by a lecture every fortnight at RAAF Fairbairn with a monthly visit away. Some of the monthly visits were coordinated with other State's IMC groups which often made for lively encounters.

Our course leader was retired Admiral Graham. He was a very pleasant man with two pet aversions – not wearing one's Name Tally (Name Plate) and being adrift (late). The former did not have too many consequences but the latter did. If transport was scheduled to depart at a specific time, it did with any consequential transport expense borne by the offending student (s).

I cannot remember all the students but there was a good mix of affable and competent representatives of the three Services, industry and the CPS. The senior student was Air Commodore Lessels who was a civil engineer responsible for all airfield construction and maintenance. The senior Army man was Colonel (later Brigadier) Ian Meibusch well known for his passionate commitment to Rugby. Other characters were Peter Ryan, Operations Manager for TAA, Commander Rolle Waddell-Wood, Fleet Air Arm pilot and Jack Doyle, a geographer from Defence Intelligence. We had a very young senior executive from BHP, whose name escapes me, but who travelled up from Port Kembla to attend the course.

Peter Ryan was a great raconteur and humorist who kept us

entertained whatever the circumstances. His forte was making a speech 'on behalf of the visiting bowlers' whenever we happened to be in a restaurant where another group was conducting a farewell. This seemed to happen on most visits away and was usually the highlight of the trip.

Early in the course we had a scheduled visit to BHP, Port Kemble. The Army was going through one of its austerity drives on travel allowance and, initially, Army students were told we could not participate. This caused a 'bolshie' reaction but the problem was solved by the Army students driving to the coast in a staff car while BHP offered accommodation in their coastal Guest House. This turned out to be quite luxurious with staff to burn and a very well stocked bar so we had the last laugh when the other students, housed in a mediocre motel, chided us on being unable to be with them.

We had one really memorable visit to Tasmania some months later when the Army travel budget was in better order. Our first call was Hobart where we visited, inter alia, the aluminium smelter before booking in to the Hobart Casino hotel complex. I used the visit to have dinner with the resident Signals officer while most of the course had dinner in a local restaurant. From all accounts this was a pretty boisterous affair aided by consumption of good wine. I returned rather late at night to find two of my colleagues sitting silently in two chairs at the entrance of the Casino. I enquired what they were doing and was told they were conducting a silent protest. On further enquiries, it transpired that these two had decided to have one more go at the tables after the aforementioned boisterous dinner. As they entered the gaming room, one of them tripped and fell to the floor. As he lay prostrate, a bouncer enquired *"Sir, what are you doing?"* The response *"Inspecting the carpet for wear"* did not impress the bouncer who then asked him to leave. The second member of the party then said *"You can't do that, he is a senior government official"*. This cut no ice and both were ejected hence my finding them conducting their silent protest. I eventually got them to bed but we all then overslept and were adrift which necessitated an expensive taxi ride to catch our colleagues at Hobart Airport to catch our short flight to Launceston.

From Launceston we drove by bus to the Defence Food Research Establishment at Scotsdale. We were not a well group with recovery from the previous evenings festivities not being helped by a very rough plane ride over the mountains.

Our morning was spent sitting in an auditorium being subjected to a number of detailed briefings about food research, the intricacies of food dehydration, the physiological demands of military activity, etc.

The tone was set by the initial briefing by the newly appointed Director of the Establishment. After a lengthy address he asked for questions. The first was *"Did the Director think that the Establishment's isolation affected its research?"*. The Director missed the point and said the Establishment was not isolated and that were good schools and other facilities nearby. When the questioner clarified by emphasising that he meant isolation from other academic and research facilities, the Director answered *"Not at all and the airport at Launceston is close by so its easy to visit other facilities"*. He then turned to his deputy, who had been an unsuccessful applicant for the job, and sought agreement. The Deputy responded in a fit of pique *"No. You are the only B.....D who gets to travel"* and returned to sulky silence.

Our next briefing was on the theory and practice of dehydrating food. He spoke like Peter Sellers saying:

"First ve are cooking the food. Then ve are drying the food. Then ve are packaging the food" followed by much technical explanation in the sing-song voice typical of the subcontinent. I thought we were all remarkably restrained and not a giggle was heard.

The final talk for the morning was given by the resident physiologist who managed to mumble through his address without moving has lips and followed his dissertation by jumping on an exercise bike and peddling like mad. We never did find out what the aim of that particular talk was about.

I'm afraid by then we had become a bit restive and mumbling amongst the ranks could be heard. I think the staff were not really impressed with us but they got their revenge by serving us reconstituted beef and dehydrated vegetables for lunch.

After lunch we returned to the laboratory and donned paper hats and white coats for a tour of inspection. About half way through, we came to the laboratory of the Indian dehydration expert who showed us a tall specimen jar holding a single large carrot. He proudly stated that the carrot was ten years old. Someone asked 'how old is the carrot" and on the affirmative response, the Goon Show fans present started to sing 'Happy Birthday to You' which completely stumped our tame Indian.

Notwithstanding all this I can honestly say that the IMC Course was very worthwhile and certainly worth the effort.

As the year drew to a close, I began to wonder if the promise made by Colonel Burnard that I would be posted to Australia House, London, would come to pass. There had been two Directors since he relinquished the appointment so I would not have been greatly shocked if some other posting emerged. While wondering along these lines I got a call from Graeme Burgess who was now the Assistant Military Secretary (Lieutenant Colonel and above) telling me that I was to be posted to the Military Secretary's office as his replacement and invited me over to see my new office. I was not overly surprised with the proposed change of posting but, as I had been to Graeme's office many times, I was a little intrigued with the invitation. Anyway, I was met by Graeme who walked me down a corridor and flung open a door to reveal a child's desk with a child's chair and a miniature hat-rack. While I stood there with open mouth, the Military Secretary himself appeared and threw a single Huega carpet tile into the office and said *"Here's your wall-to-wall carpeting"*. I was suddenly surrounded by a number of colleagues who had a good laugh at my expense. I thought it was a good joke which was much improved when Graeme handed me my posting order to London.

As the year progressed, there was much political tension and stories abounded that the government would soon run out of money and would be unable to pay the salaries of public servants and the military. There was some evidence that these stories were being promoted by the government in an attempt to make the Opposition grant supply. Against this background, we had reached the culmination of a year of celebrations to mark the 50th Anniversary of the granting of the title Royal to the Corps of Signals. The weekend of the 8th and 9th of November was designated as the Corps Weekend with the Governor-General Sir William Kerr being Guest of Honour at the Jubilee Parade, the Garden Party that followed and later in the day, the opening of the Corps Museum.

During the Jubilee Parade the Governor-General addressed the assembly and after the expected words of congratulation, launched into a message on duty, the Constitution and various other matters which left most of us convinced that something dramatic would follow within a matter of days. There would have been very few present on that day that would have been taken by surprise by the Dismissal.

During the Garden Party that followed the Parade a number of us were attending the Governor-General when he was asked if he would care for a drink. He immediately responded with *"A gin and tonic please – I always drink G and T in public as it looks like lemonade"*.

After the Garden Party we repaired to the Museum for the opening ceremony hosted by Brigadier Keith Colwill who was the Representative Colonel Commandant at the time. When all were assembled a team of runners from 139 Signal Squadron led by 2nd Lieutenant Challis mounted the steps with a Jubilee Message which had been carried by a relay all the way from Townsville – a quite remarkable feat. The Jubilee message was handed to the Governor-General who duly declared the Museum open and then inspected the collection.

The Museum itself had a rather chequered life and was relocated a few years later and reopened on at least two other occasions – but more on that later.

Back in Canberra I was soon engrossed in handing over my appointment and attending briefings for my new one. The London job was a dual one – Staff Officer Communications and Defence Communications Liaison Officer. Although my posting was not as an Attaché, I still needed to be briefed by the various intelligence agencies, the Joint Communications Branch and Material Branch. It was a hectic time and I was looking forward to the challenges of the dual appointment.

I must say that I had been some what surprised when my family did not share my enthusiasm for the posting to London. I had thought this to be a great opportunity for the children to see Europe and had sought the posting, possibly to the detriment of my career, as a unit command would probably have been more advantageous. In retrospect, it might have been better if we had stayed at home as home life had become somewhat strained.

While I had been fully occupied with the personnel and domestic side of the Corps, much had been going on with the rationalization of fixed communications. As the developments would lead to my involvement while in the UK, some background may assist the reader.

Up until 1970, each of the Services had maintained their own communications networks providing secure telegraph messaging. Extensive use was made of Post Master General telegraph lines all augmented by separate high frequency radio networks. There was considerable waste within the systems and, in many instances, transmitting and receiving stations for each of the Services were almost colocated. Another driving issue was the growing demand for secure high quality speech, secure facsimile and data along the lines of what had been proposed for MALLARD. Army had an automated system for switching telegraph messages (STRAD) but the other Services were sadly lacking in this area.

In 1971 a Defence Communications Rationalization Study (DCRS) was initiated with a report due for delivery in 1973. The major recommendations of the DCRS Steering Committee were:

- The formation of the Defence Communications Division within the department of defence to manage the planning, procurement, installation and operation and maintenance of the fixed communications network;
- That the system provide all modes of communication, with the possible exception of video, within the same network; and
- That, during the transitional phase to the new network, there would be ongoing rationalization of the existing networks.

In January 1976, the Defence Communications System (DCS) Division was formed under a General Manager with two star status. The GMDCS was to be filled by either a military officer or civilian with the first appointee being Mr Ian Maggs. At the same time, the Defence Communication System User Committee was established to provide user advice to the GM. I anticipated, correctly, that I would become heavily involved with the Ministry of Defence and UK industry as the project, now named as the Defence Integrated Secure Communications Network (DISCON), developed.

To return to my story, I was due to arrive in London just before Christmas to take over from my old friend, Keith Carey, so planning started in earnest to arrange for the trip, pre-purchase a duty-free car, sell cars and arrange for renting the house, etc. I thought it would be a good idea to take some leave on the outward journey and have a few stopovers to appease the family.

Our first stop was Singapore and we spent a day there touring the island in a taxi I had hired for the day. The next stop was Bangkok where we were to join a flight to Athens for a two day stopover.

Our first travel glitch occurred at Bangkok when the late arrival of our flight meant that there was not enough time for our luggage to be transferred. I was given two options-proceed as planned with luggage to follow on the next flight or wait until we could travel with our luggage. I was set to opt for the second alternative when we found that the Bay City Rollers would be on the earlier flight. As Kerry was an avid fan, the decision was made for me and Kerry spent a number of rapturous hours with her idols.

As might have been expected our luggage did not appear on the next flight as promised so I had to spend a lot of time visiting and revisiting the airport and personally searching baggage rooms. I did eventually find the missing items but it did cut into our tourist activities although we did see the major sites in Athens and did take a short cruise.

The hotel where we stayed was a popular tourist hotel much used by Americans. While there we were entertained by an American tourist decked out in plaid shirt, non-matching plaid shorts with sneakers and black sox proclaiming that the sites of Athens were just a bunch of god-dammed rocks, no-doubt brought from all over the country and dumped in a few strategic positions!

Next step was Rome for two days. My family did not share my enthusiasm for the Eternal City and did not appreciate being dragged from pillar to post to see as many of the famous sites as was possible to see during our short visit.

We flew to London with Air Italia. We were in First Class with a clear view of the cockpit and were able to witness an argument between the Pilot and Copilot which went on for most of the trip and did nothing for our confidence when we ran into bad weather as we approached the UK.

The second major glitch now occurred as we were unable to land at Heathrow and after much circling around were diverted to Luton Airport. Because of the diversion there was no one to meet us and I was forced to hire two cabs to take us and our luggage to London. It was a really inauspicious start to our two year posting.

LONDON

Let me deal first with the domestic arrangements. Rented accommodation was at a premium and most members of staff lived on the outskirts of London. After looking at a few places in Kent, we settled on taking over the Carey's house in Walton-on-Thames. This was very much in the stock broker belt with some high quality rentals and a good train service to the city. It was a good service although often disrupted by strikes and breakdowns. What I found interesting was the fact that, almost irrespective of where one lived, it took an hour to get to work. This was certainly the case for Jim Farry and me – Walton was served by a train which had multiple stops while the train from Woking, where Jim Farry lived, was an express service – the overall result being a two hours commute each day.

Keith Carey was due to depart the UK on 7 January which gave us little time, not helped by the Christmas holiday period, for a hand-over of duties and contacts. Our fist visit was to the Royal Aircraft Establishment, Farnborough, a short train journey from London. I cannot remember why this visit was necessary but it ended up by giving me a quick insight to the Irish terrorist problem and the general nervousness of the population. We arrived at the Farnborough Railway Station to find the platform exits blocked and a reasonably large presence of police and bomb squad personnel. An unattended large white Samsonite suitcase had been left on the platform which had triggered the bomb squad response. Passengers were unable to leave the platform but we were herded away from the target suitcase and in due course a soldier with extensive padding fired a number of well aimed shots. There was no explosion but the suitcase was shattered and piles of clothing were scattered all over the place. The whole performance took an hour or so and we returned back to London. This was not an unusual occurrence nor was the practice of London commuters throwing unattended briefcases from trains.

While Keith was still on hand we travelled to the Government Communications Headquarters, Cheltenham, to establish some contacts and for me to establish my credentials and access level. Keith also accompanied me on visits to the Ministry of Defence to call on the Signals Officer-in-Chief and various other staff directorates. After receiving a few more words of wisdom and a good contact list Keith was off and I was on my own.

In those days there was a large Defence presence at Australia House. There was a two star Head of Australian Defence Staff with each of the three Services having a one star. The Australian Army Representative was Brigadier (later Major General) Ron Grey. I had known Ron for many years but we had never served together. I had met him just before leaving Australia and I expected a friendly reception. Thus, I was more than a little surprised when he told me at our first meeting that I could forget being his Staff Officer Grade 1 (Communications) as I was to be his Staff Officer Grade 1 (Operations). I did not quite understand this due to the nature of our work in the UK. I have to say that for the first few months I was at a loss and did not really know where I stood with him. However, at an early stage in my appointment he gave me a directive to conduct a study on the equipment and tactics of Mechanised Infantry as practised by UK, the USA and Germany. I was not well equipped for this task but the report must have been to his liking and I became 'one of his boys'. This meant that I could do no wrong and had almost unfettered access to travel. It also meant that over the years that followed I had a champion who went to bat for me on numerous occasions. We became and remain good friends.

Back to the domestic side, we had settled well into Walton-on-Thames and the children were enrolled in good schools. The SAAB that I had ordered duty free from Sweden finally arrived so we were able to see something of the countryside on weekends. The local travelling did not last long as the children soon established a circle of friends and were reluctant to go on weekend trips.

When we arrived in London it was overcast and it seemed to remain so for months on end. I was going to and from work in the dark which became very wearing. Yvonne got herself the first of a few jobs and seemed to have settled in well.

My office in Australia House was on the Aldwych side which gave me a good view of the senior 'city' folk who appeared to come to work at about 10:00 in the obligatory Roller, head to lunch at 13:00, return from lunch at 15:00 and depart the city at 18:00. I still do not know how they managed to keep the wheels of industry rolling but they obviously did.

My responsibilities included the Communication Centre on the top floor of Australia House. This task was not particularly onerous as I had a good team and I usually only got involved when messages were over classified and I had to bring this to the attention of the culprits.

Each of the Service staffs maintained a small 'Mess' where we kept an adequate supply of Australian beer and wine for our own use and for entertainment purposes. This was extremely useful so that we could provide return hospitality for our industry contacts and to 'grease the wheels' in our workings with military counterparts.

Apart from the Mess, I also found membership of one of the military clubs to be very useful. I was a member of the Naval and Military Club, Melbourne, and expected that I could make use of the reciprocal clubs in London for the duration of my posting. I soon found that this was not to be the case as the Naval and Military Club (The In and Out) and the more prestigious Army and Navy Club would only extend reciprocal facilities for six months after which a nomination fee and full membership fees would be required. In contrast the RAF Club was happy to oblige as long as I remained a financial member in Melbourne.

Most of the Army staff were also members of the Wig and Pen Club in the Strand. Originally built as the home of the gatekeeper of Temple Bar, the Wig and Pen is said to be the only structure in the Strand to survive the 1666 Great Fire of London. The attraction for the staff was our easy access to it when Australia House was closed as the result of a bomb threat which occurred reasonably often. A threat, usually received by telephone from a caller with a broad Irish accent, was always acted on and staff were required to clear the House and to assemble outside St Martin in the Fields just opposite our main door. The House had to be cleared for at least two hours and the bar of the Wig and Pen was far more congenial than the steps of the aforementioned church.

The military staff at Australia House officially worked the same hours as we would at home but most of us found we had to stay late to keep up with our workloads. Travelling time to and from work meant that any after-hours commitments such as High Commission, Diplomatic or other functions could only be met by remaining in the office. The IRA threat precluded wearing of uniforms on public transport so we all found it necessary to maintain an extensive wardrobe in our offices.

Because of the 'no uniform in public' policy, it was necessary to possess additional civilian suits. I tried the usual offerings of Burtons and similar but found them not completely satisfactory. I'm not sure where the referral came from but I became a customer of "arry Levy, bespoke tailor of "ammersmiff' which translates to Harry Levy of Hammersmith. Harry had moved from Saville Row to reduce his overheads and he produced excellent clothes. A few of his suits lasted me for many years.

My predecessor, Keith Carey, had been mainly involved with the MOD and Royal Signals. He had been very effective in establishing close liaison with influential members of MOD, the School of Signals and with various Royal Signals units. He had made many good contacts with industry but this was secondary to his main work. By the time I had replaced him, the emphasis had changed and, although I maintained close relationships with the contacts he had made, I had been directed more towards industry. My brief for the appointment gave me many lines of enquiry to follow in support of the projects of interest but I also had some leeway to research other areas of technology at my discretion.

The span of Australia's interests in the communications field were extensive and growing. I was tasked with not only gathering information on the military user requirements side but also making contacts with the main players in industry. Australian communications projects included DISCON; high powered HF Radio equipment for the fixed network including the transportable systems (Hi Port, Med Port); Combat Net Radio (RAVEN); and area communications systems (this would eventually become PARAKEET) to fill the gap left by the cancellation of MALLARD.

As the year progressed and after a visit by Lieutenant Colonels Catanarch (the Australian Army EW guru) and Dick Twiss, I was also tasked with becoming knowledgeable on Electronic Warfare developments in both the passive (Electronic Support Measures) and active (Electronic Counter Measures) fields. To fulfil the EW role I had to make additional contacts in specialist areas of Siemens, Thompson CSF, Telefunken and Plath. Some aspects of EW were new territory for me and I found the work intriguing to say the least.

At this time in the UK, there was feverish activity within industry to respond to the requirements of the Defence Communications Network (DCN); establishment of an extensive satellite communications network; development of a complete range of combat net radio to be called CLANSMAN; and a digital area communications system,

PTARMIGAN, to replace the ageing analogue system, BRUIN, which was the mainstay of the British Army on the Rhine (BAOR). There were a myriad of players in the field including such UK firms as Plessey, Racal, STC, Cossor, British Aerospace and Muirhead; Thompson CSF in France; Siemens in Germany; and a Norwegian firm whose name escapes me.

My first contact with industry was with Air Commodore Paddy Fagan from Plessey. Paddy had had a long career as a pilot and had flown in the Royal Navy Fleet Air Arm before continuing with the RAF. He was an absolutely charming gentleman who was a regular guest in our Mess. Given half a chance he would regale us with stories of his life in India which, despite regular repetition, were very enjoyable. Anyway Paddy was instrumental in my making contact with the hierarchy of Plessey at both the technical and engineering levels.

As I was establishing contacts in industry, I was also going through the process of obtaining the necessary passes for entry into the MOD and meeting officers in the office of the Signal Officer in Chief (SOIC) and other areas that dealt with intelligence and antiterrorism. All was not plain sailing in getting the access to appropriate people until I remembered advice I had received from Keith Carey about ensuring I brought my Staff College tie with me. I didn't quite believe Keith until I tried wearing it and found all sorts of doors opening. I understand that this was due to the high regard given to staff college training and the status the tie brought with it. By the end of my tour I had worn out one tie and was on to my second one!

One of my early trips outside London was a return to GCHQ for more extensive briefings including a refresher course on Signals Intelligence with an emphasis on the work done at Bletchley Park during the war. I was aware of some of the achievements at Bletchley Park but was amazed at what I was told at the briefing. All of this was highly classified at the time but much has been revealed in recent times. While at GCHQ I was made aware of the work being done at their affiliated Speech Research Establishment but more on this later.

Early in the year I also attended the Headquarters of the Defence

Communications Network which was very relevant to my work for DISCON. Later in the year I visited an underground communications centre that was housed in the depths of an old mine. Before descending, visitors were given a metal disc to put in one's pocket to enable identification in the event that a fire broke out. A little scary!

It was not long before I realised what Brigadier Grey had meant when he said I was to be his Staff Officer (Operations). Because of the various Security Clearances I held as a Signals officer, I was in a good position to deal with certain aspects relating to the Special Air Service. For reasons that I never quite understood, Brigadier Grey had been directed by Australian Army Headquarters not to get involved with anti-terrorism. With great foresight, he chose to ignore this and I was despatched to SAS, Hereford, to establish contacts and become conversant with all aspects of their anti-terrorism capabilities, which were formidable indeed. I had ongoing involvement in this area which was extremely valuable in later years when I was Director of Communications and my classmate, Mike Jeffery, was the Director of Special Action Forces.

* * *

I had been advised soon after arriving in London that it was not always easy to take leave and that short bursts were preferable to a long break. Early April was a quiet time so I took the family to Eire for a week. I chose Eire as our first trip as there was a family connection on my mother's maternal side and a number of the Brothers who had taught me were living there in retirement. I rented a thatched cottage in a town called Holy Cross near Thurles in County Tipperary. My old school had been named after the Cistercian Abbey of Holy Cross which had been destroyed in Cromwell's time but was being restored when we visited.

It was not possible to make the journey from Walton to Holy Cross in a day so I sought advice from the older hands at the High Commission. All recommended Mrs McQuade's Guesthouse not far from the ferry terminal in Rosslare. I did not receive precise directions but was told to drive towards Killinick and then ask a local.

I did as I was told and stopped a likely looking local. He was the first of many caricatures of Irish men that we met on the trip. His response to my enquiry was along the lines:

"Certainly Sir, I know the place well. Take the second left along this main road. On the top of the hill you will see a big white house. Ignore that!" and so on and so on but ending with *"You can't miss it"*. Well, miss it I did and we ended up staying in a hotel along the way.

On arrival in Holy Cross, we were told that the cottage was not quite ready for us and we should spend the time driving in to Thurles where an All Ireland Hurling final was to take place. I was glad we did as it gave me a topic of conversation when later that day I visited the local pub at Holy Cross.

The pub was straight out of the old John Wayne and Maureen O'Hara movie 'The Quite Man' complete with the parish priest sitting quietly in a corner and a patron who, when bored, wandered through the pub asking patrons if they would like a fight. The pub had a notice above the door which stated 'We're Open 'till We Close'. At first, I was regarded with some suspicion so I very quickly identified myself as an Australian who had been educated by the Brothers of the Order of St Patrick. The final barrier was broken down when I correctly identified a miner's carbide helmet lamp. Once accepted, I found it necessary to be in attendance at 17:00 otherwise they would send out a search party.

There was always traditional music in the pub and I spent many happy hours there. As indicated by the pub sign, operating hours were very flexible. The publican had a most effective way of closing the place down when he felt the need. He would simply order the musicians to play the national anthem. All the patrons would jump to their feet in respect for the anthem, then quickly finish their drinks and leave without demur.

I soon located Brother John Gallagher and arranged to collect him from a town a little South of Holy Cross, then collect the family and proceed to the premier school of the Order at Ballyfin. John and I were passing through a small village en route to collect the family when he suddenly called out *"Stop the Car"*. I did as bid in some alarm and asked what the problem was. In a much quieter voice he said *"It's 10 o'clock – time for a drink"*. Into the pub we went for a Guinness and an Irish whiskey to 'wet the glass'.

We got to the school to find Brothers Pious and Majellan waiting for us. We were royally entertained and had a grand time reminiscing. I have a vivid memory of my daughter Kerry, who was assisting serving drinks, asking Brother Pious what he would like with his whiskey and his response *"a little more whiskey"*.

We all enjoyed a great day but I was saddened by the fact that my children did not understand why I would want to call on my old teachers.

The week was soon over so it was back to the ferry and then home.

* * *

In addition to my normal tasks, I was also called upon to represent Australia in various multinational panels dealing with various aspects of communications standards and interoperability. These were always enjoyable affairs and I particularly liked catching up with colleagues from the USA and Canada. One of the more interesting meetings I attended gave a great example of the Churchillian statement that the USA and UK were separated by a common language. We had been discussing a particularly vexed issue on interoperability and were making little headway The UK had come to the meeting wishing to table a paper on the subject. This lead to heated discussion for a day and a half until the USA delegation realized that tabling meant the Brits wanted to open the paper for discussion whereas the USA thought the UK wanted to put the paper under the table and defer discussion. Once the language barrier was broached agreement was speedily reached.

Apart from the usual involvement with the MOD and industry, there was always something going on to make the UK posting very enjoyable. Of the many interesting events in 1976, two stood out in particular, these being a visit to the Chelsea Hospital for the Founders

Day Parade and some weeks later the Trooping of the Colour. The former was a grand event with the Chelsea pensioners in their red uniforms circling their parade ground at their special march pace, somewhere between Slow and Quick Time. They were an impressive lot with medals galore and twinkling eyes. After the parade there was the Founders Day Remembrance Service in the Pensioners' Chapel. The Pensioners entered first and once they were settled the guests processed in with suitable musical accompaniment. As we went down the aisle, to my amazement I felt Yvonne flinch a few times. When we were seated she whispered that her bottom had been pinched by a number of the young at heart old men! I heard later that no woman was safe in the Chelsea Hospital area from their attentions, not always unwanted.

The Trooping was a magnificent sight, particularly from our vantage point of reserved seating. I had seen films of the Trooping many times but none of these ever successfully conveyed the pomp and ceremony and the excellence of the music.

My mother arrived in the UK in mid-May for a two month visit. After a few days she advised me that Bert Fry, a widower that lived a few doors away from our family home in Ryde, had asked to marry her and she sought my reaction. I thought it was a great idea and was pleased that she had found someone.

During her stay she had a trip to Paris with Yvonne and later all five of us went to Cornwall for a weekend away. The hotel was right out of Faulty Towers. When our car pulled up at the hotel a man in a grey dust coat collected our luggage and took it to the reception desk. He appeared almost immediately again wearing a dark jacket and signed us in. Later on he appeared in the bar wearing a white jacket to serve drinks while later again he served our meal wearing a dinner jacket.

In June we had a ten-day visit by the Chief of the General Staff, Lieutenant General AL McDonald, a rather humourless and formidable man. Planning for the visit had been extensive and painstaking. Courtesy calls on important people had been set in place, a reception was organised at the Tower of London and VIP arrangements for attendance at two major Army exhibitions had been organised.

Brigadier Grey planed to met the CGS at Heathrow at the aircraft steps and usher him personally through Customs and Immigration. Two Daimler Limousines were ordered to ensure a back-up was in place.

We thought all bases had been covered but then Murphy's Law kicked in with a vengeance. First, both Daimlers broke down and by the time Brigadier Grey got to Heathrow, the CGS was sitting in an airport bus. The next hiccup occurred at one of the Army equipment exhibitions where the CGS's name had been left off the VIP List and he and the Brigadier were placed with the hoi polloi. When things looked as if they could not get worse, a helicopter ordered to get the CGS and party back to London for the Tower of London Reception went to the wrong rendezvous point leading to a very rushed return journey. From memory, there were a few other mishaps so we had on our hands a CGS suffering an extreme sense of humour loss who was giving our Brigadier a hard time.

Although not taking direct part in the CGS visit programme, we three Lieutenant Colonels, Warwick Smith, Jim Farry and myself, were required to be in attendance at Australia House at the end of each working day to, respectively, effect any necessary administrative requirements, answer technical questions and handle any communications. The CGS was quite pleasant to us but continued to address the Brigadier simply as '*Grey*' and generally give him the Fourth Class treatment. As soon as the CGS was off our hands, we three Lieutenant Colonels gathered with the Brigadier in our Mess to try to quieten him down. When we asked what the Fourth Classing was all about, the Brigadier told us the story but that is his story not mine so I wont repeat it.

The day following the departure of the CGS, I departed London for the first of many trips to Paris (Thompson CSF), Ulm (Telefunken) and Munich (Siemens). As time progressed I visited other firms and locations but these three remained my main area of interest in Europe. Rather than attempt to relate my many visits in a chronological order which would require a feat of memory beyond my capabilities, I will just mention some highlights.

I visited Siemens more than any other firm so I will start there.

Siemens could have bid for any of the major projects mentioned earlier so I had many visits to Munich and other centres of manufacture as well as R&D.

I had a particular interest in the work being done on high frequency radio, particularly with regard to data transmission. HF Radio was the mainstay of long distance communications and would still have an important place to fill even after the introduction of satellite communications. HF radio was generally used for analogue insecure speech and multichannel telegraphy. It was possible to transmit low rate data at about 2.4 Kbps but this was not always reliable and error rates could be quite large. The approach being researched at Siemens was based on calculating the waveform that would be generated by various signals, synthesising the waveform and then producing a transmitted signal via a wide band linear amplifier. It was an intriguing approach and early research results were very promising. My interest in this area increased when I became involved with Vocoders – but more on this later.

My first visit to Munich was very interesting. I was met at the airport by one of Siemens' marketing staff who settled me in my hotel and then proposed a stroll through the city centre before dinner. My guide was very affable and knowledgeable about the city and we were getting on splendidly until we came to Munich Cathedral. My guide insisted I look at the photographs of the World War II destruction and waxed lyrical about the barbaric bombing. I then simply asked *"Have you been to Coventry?"*. Our evening then ended abruptly and I was escorted, in silence, back to my hotel. I'm not sure how this was reported to Siemens management but I never saw the man again.

The next day I was met by an engineer, Erwin Schäuble, an absolutely charming man with impeccable manners and a very good command of the English language. He was to be my escort on this and every subsequent visit to Siemens and we became good friends.

When we arrived at the Siemens Research Laboratory, I was introduced to the HF Radio research team. I cannot remember the leader's name but I have a clear recollection of every question of mine being answered with the preamble *"Let us first look at the Physics"*. This would be followed by detailed descriptions and mathematical derivations. I was fascinated by the work being done in this field but was not particularly successful in transferring my enthusiasm to the appropriate Australia authorities.

One of the more notable aspects on my dealing with German firms was the efforts they made to conduct our discussions in English. When it became necessary for them to revert to German for an obtuse point, they would apologise, have the discussion and then attempt to translate whatever was said. This was in marked contrast to the French who clearly thought that anyone who did not speak French to be a barbarian and should be treated accordingly.

I always enjoyed my visits to Germany but my favourite city was Munich and its surrounds. In a day it was possible to visit the Palace at Herrenchiemsee, visit a range of fascinating museums, walk through Marienplatz to enjoy the restored city, visit the surrounding mountains and finish the day in one of many raucous beer halls or gardens. My favourite was the Mathäser, the largest in Munich with a capacity of 5000, where I spent many a pleasant evening after work.

As an aside, I found the German attitude to work and play very interesting. Work was work and they were assiduous in giving their all during the working day. During my visits, I was invariably collected from my hotel early enough for an 08:00 start and I would not be released until about 18:00. They were just as keen on play which would generally involve a restaurant and a beer hall.

I had regular visits to Ulm with Telefunken. Telefunken produced a mobile Very High Frequency jammer mounted in an enormous wheeled vehicle that could set up and tear down in a matter of minutes. The technology was fascinating and incorporated scanning receivers to detect enemy transmissions and quickly switch to high energy jamming. I was hosted by Telefunken on a number of occasions to visit German Army units in Rommel-Kaserne to witness equipment demonstrations and trials.

Ulm is a beautiful city famous for the Ulm Minster, the highest church in the world. The Minster is a beautiful building and I very much enjoyed an extensive tour of it. There were a number of memorial tablets for World War II just inside the nave. The stone tablets showed no names but simply the number of soldiers killed in the various regiments with connections to the city.

My other regular destination was Paris for discussions with Thompson CSF. Thompson CSF had much of interest including the RITA Area Communications system, an excellent range of man-pack and vehicular net radios and the BROMURE EW system.

I had one memorable trip with Thompson CSF to their man-pack radio factory in Lisle. My escort and I travelled by train from Paris and I was conducted around the factory by a group of engineers. I was finding the tour extremely interesting and asked lots of questions. I noticed my escorts getting rather fidgety as midday approached and we hadn't finished. Eventually it got too much for them and I was told we just had to break for lunch. We left the factory and followed a path to what looked like a large concrete air-raid shelter. I must say I was feeling somewhat dubious when we entered until I found we were in what looked like a provincial restaurant with checked tablecloths and an extensive wine selection of wines. I had half expected a light lunch but it ended up being one of the most memorable meals I have ever eaten and it took about three hours.

After lunch, my escort and I were driven to the railway station in time to catch an express train back to Paris. We had no sooner settled in our seats when we were peremptorily told to find other seats as a Government Minister had commandeered the whole carriage!

My experiences with the French were markedly different from those I had had with the Germans. They could be very charming and helpful as individuals but collectively I often found them very difficult to deal with. When there were to be serious discussions on capabilities, etc, a government official was always present, in fact, I often found myself sitting at a conference table for extended periods waiting for some official to arrive.

I made only two or three visits to Hamburg for demonstrations of

Plath direction finding equipment. The visits were interesting and my hosts were generous in their time and hospitality but I never established anything like the close association I had with the folk at Siemens. Perhaps this was because the northern Germans are more staid than their Bavarian brothers.

As I have said, I very much enjoyed my many visits to Germany and the personal contacts that I made. The people I met with were well mannered, educated, hospitable and cultured and I must say that I had difficulty coming to grips with the fact that this same race was capable of the horrors of World War II. Many years later, I had similar thoughts after visiting Japan with a group of State RSL Presidents.

To take up the story again, my Mother left the UK in mid-July and a few weeks later the family commenced a tour of Europe. This was one of those city a day with bags out at 06:00 bus tours that covered Belgium, Austria, Germany, Italy and France in fourteen days. I had booked with an allegedly up-market company and was very disappointed with the experience. Our hotels varied from very good to almost unspeakable with one of them being infested with cockroaches. So also was the food offered on the tour. I had thought, quite reasonably, that we would sample regional delights but, in fact, we were served chicken and chips with monotonous regularity. When I complained, I was told that the menus were designed to meet the requirements of the British tourists. The trip had been less than a roaring success with the family and I was glad to get back to work.

I was away from the office for much of the next few months visiting various industrial and government R&D establishments, observing equipment trials and visiting factories. I was developing a very wide circle of contacts in government laboratories and industry, particularly with RSRE, Christchurch, and the Speech Research Laboratory in the case of the former and with Racal and Plessey in the latter.

In October 1976, I had a visit to Wiesbaden in Germany which was something out of the ordinary for me. Wiesbaden was a very important centre for the US Army in Germany and a manufacturing centre for Siemens. I have not been able to remember precisely why I went to

Wiesbaden but it was certainly in the company of Erwin Schäuble so it must have involved Siemens in some way.

Wiesbaden is one of the oldest spa towns of Europe and is famous for its hot springs. It is also famous for its casino which had been closed with all other similar gambling establishments in 1872 by the Prussian dominated Imperial government and reopened in 1949. I had never been in a casino so I thought I would try this new experience.

To enter the casino, I had to show my passport to prove that I was not a local inhabitant. I went to the cashier to buy chips and was somewhat taken aback to find the minimum was DM 10 which was close to £1 at the time. I started playing Blackjack which looked like Pontoon but was soon disabused of that idea by an irate croupier. After a string of losses I moved to the roulette wheel and wasn't doing much better. In desperation I moved to another roulette table and started betting on a block of four numbers with my single DM 10 chips while standing next to a number of players betting with handfuls of DM 1,000 plaques. My luck changed and I had a run of eight wins and suitably encouraged I placed five of my hard won chips on a single number which came up. It took all my self control to stop yelling out *"You bloody beauty"* as I had recovered all my losses and had made a substantial profit. Not wishing to press my luck, I gave the croupier a tip and departed swiftly to my hotel.

Later in the month I had a break from the normal run of things as I became involved in the transfer of Major (later Lieutenant Colonel) Mike Collins from Royal Signals to the Royal Australian Corps of Signals. Mike had been on exchange at our School of Signals and had become enamoured of Australia. We became close friends and he later became a member of my staff.

* * *

Earlier in the year after settling in to the house in Walton-on-Thames, I thought I would get myself fit by joining a local Rugby club. Not far away was Esher which had a team called the Expendables. This sounded just my speed and I had visions of gentlemanly games followed by a few beers in the clubhouse. I soon found that the minimum entrance level to the Expendables was a county cap and I ended up playing in the lowest competition grade which contained a number of hopefuls looking to make a reputation.

I played one game in which I cracked three ribs in the first 10 minutes which probably sets some sort of record for the shortest Rugby career ever. Anyway, I attended at the surgery in London of the retired RAMC Colonel who looked after the colonials. As I entered his room all bent over, he jumped to his feet exclaiming *"My dear chap, have you had a car accident?"* When I replied I had been playing Ruby his response was *"Bloody idiot"* and sent me to be suitably strapped by his nurse.

Although my playing career was over, not so my interest in the game. I did manage to get to Twickenham a few times including once when Ian Meibusch was visiting from Australia. I was fascinated by the fact that, although enormous quantities of beer was consumed by the crowd, there was almost a complete absence of crowd misbehaviour. This was in stark contrast to crowd behaviour at major Soccer games which led me to ban the children from attending such events.

I also had the very good fortune to be invited to Wales with Brigadier Grey and a few other members of staff to watch an England versus Wales game at Cardiff Arms Park. This really was one of the highlights of my posting to the UK and I have vivid memories of being in the stand listening to the Welsh sing and feeling the hair on the back of my neck rise. Another side benefit of this trip was to be introduced to the work of the Welsh comedian and folk singer Max Boyce which continues to entertain me.

Having decided that Rugby was not an option, I joined the Walton-on-Thames Hockey Club at the start of the 1976 winter season. Damian and Kerry were good players and they joined with me. I must admit this was much more fun than Rugby especially when I found that I had retained my skills from RMC and was soon playing as either wing or centre – forward with the men's team.

There was no fixed competition system of matches in local hockey so one member was deputed to arrange home and away games. There was a simple formula – drive to the club or to an away venue, play the game and then enjoy high tea. I played with the men on Saturday whenever I was able and often with a mixed team on Sunday. I soon found that playing against females was very dangerous indeed and decided a single game per week was adequate.

I was very much enjoying my hockey and was scoring more than my share of goals until one afternoon when on the wing I fell to the ground in great pain and thought I had been hit on the calf with a stick. I tried to get to my feet to look for the culprit to find there was no one in the vicinity and I had difficulty standing. The following day I attended the local GP's surgery who told me I had torn a calf muscle and nothing could be done and that I should expect to be using a stick for three or four months. I didn't find this approach very helpful so I got myself sent off to the Queen Elizabeth 11 Military Hospital. They had an extensive sports medicine centre and I soon found myself treated with ultrasonic massage, hot and cold compresses and was fitted with a complex elastic bandage that held the muscle in place. This excellent treatment had me off the stick and right as rain in just over a month.

*　*　*

During the year there were many visits from politicians and senior government officials. Mostly these did not concern me unless special communications arrangements were necessary. I continued to be heavily committed with visits to industry, either following my own lines of enquiry or escorting Army Office or Defence visitors.

We rounded off 1976 by dining out Brigadier Grey who was staying in the UK to attend the Royal College of Defence Studies, a pretty certain indicator that promotion would follow. The new Australian Army representative was Brigadier Geoff Leary. We also had a new Head of Defence Staff, Admiral Robertson, with his new staff officer, Lieutenant Colonel John (Bones) Bertram.

1977 was very much like 1976 except that I found I was spending more time with Royal Signals and made good contacts with the Signals Officer in Chief, serving officers as well as with the Master of Signals (our equivalent is the Representative Colonel Commandant), the various Colonels Commandant and the Royal Signals Association. I was most impressed with the work done by their Corps Committee in their charitable works as well as their commitment to heritage exemplified by their ongoing acquisition of suitable works of art and Corps silver. The Corps Committee also produced a newsletter called 'The Wire' plus a more academic offering 'The Signals Journal'.

Through the Signals Officer in Chief, I was able to visit 1 British Corps of British Army on the Rhine as an observer of the annual command and control exercise, which in 1977 was known as Fighting Falcon. In the UK system, communications and headquarters support staff were combined so within the Corps there were a Corps Headquarters and Signal Regiment, three Divisional Headquarters and Signal Regiments plus a variety of other units. I was able to observe at first hand the operation of a highly mobile area communications system at a level that we would probably never see in Australia.

One memorable experience was my visit to the Corps Headquarters and Signal Regiment. I had been travelling around the communications area with the Commanding Officer and asked him how he went about choosing a suitable site for the Corps Headquarters. I was expecting a response that would include the technical requirements necessary to ensure communications together with an expose on the tactical considerations for protection, ease of access, setup and tear down, etc. In response, he gave me a rather withering glance that indicated he thought my question verged on the absurd and simply said:

"I find a suitable place to commandeer for the Corps Commander's Mess and we set up around that!"

On a separate trip to Germany, I visited two Royal Signals EW Regiments that hosted a number of Australians on exchange, including my classmate, Steve Hart. At one of these I was shown the decrypt of a Russian signal that indicated the unit I was visiting was a top priority for destruction if a shooting war started. At this same unit, I observed one of the Warrant Officer cryptanalysts doing the *Times* crossword. He placed the crossword on his desk and walked up and down a few times glancing at the puzzle as he passed before sitting down and completing the whole thing in one fell swoop in a matter of a few minutes. I gather he thought that writing in individual clue solutions was cheating as it gave hints on the other clues. He was straight out of Bletchley Park!

*   *   *

During 1977 my tasking in support of particular projects did not change although the emphasis did from time to time. I spent much of my time out of the office travelling extensively in the UK and Europe. With regard to the latter, I was still making regular visits to Thompson CSF, Telefunken and Siemens with occasional visits to other communications-electronics firms.

In response to a particular request from the Electronic Warfare staff in Australia, I made contact with the Italian firm, Electronica. They produced a very elegant ESM suite of equipment in a portable shelter. As part of my visit to Electronica, I was sponsored down to Anzio to visit the Italian Army Electronic Warfare School to see a demonstration of their equipment.

My escort and I drove down the coast from Rome and I was intrigued to see the a great number of eucalyptus trees, very reminiscent of home. We arrived at the School at an inconvenient time, afternoon siesta following lunch, and had to wait outside the gates which I must admit I found irritating.

Eventually, my escort and I were met by three immaculately tailored Italian Army officers wearing leather gloves and pointed shoes. As I said, they were immaculate but some of the soldiers I saw were scruffy to say the least and many were wearing love beads. I was given a briefing on the Electronica equipment in a mixture of English and Italian and was then ushered into what was a rather small shelter for a demonstration. The demonstration could only be described as a shambles as it was given by all three officers simultaneously with each of them trying to twiddle the same knobs and all getting rather excited. I must say it was a relief when it ended and my escort and I escaped back to Rome.

* * *

Midyear in London was exciting as we had the Royal procession to mark the Queen's Silver Jubilee. This was a grand affair and we on the staff at Australia House had a great vantage point to view it. This was also the season for the Trooping of the Colour and for the Royal Garden Parties. Yvonne and I were privileged to be included on the list for presentation to the Queen and Prince Phillip during one of these parties which meant that I had to purchase a new Herbert Johnson cap and attend briefings on what to do and how to speak at the 'presentation'.

One of the major events of the Silver Jubilee was the Review of the Fleet at Spithead. A number of RAN ships were to take part and there was much activity in the HADS office with a particular amount of effort being made for a RAN Cocktail Party. I only became involved in this as there was a spate of 'Exclusive For' messages going to and from Australia which often meant my Communications Centre staff being called in after hours to perform the decryption. When I discovered that most of these signals were only concerned with the proposed Guest List for the Cocktail Party I remonstrated with our Admiral and got a thick ear for my trouble. It seemed that the possibility of the List leaking out could be a cause of great embarrassment and, therefore, required a level of security classification generally reserved for national emergencies.

There was a Royal Tournament in July and Yvonne and I were fortunate enough to receive an invitation. The Tournament had to be seen to be believed with the highlight being a relay race which involved four teams of sailors disassembling a field gun, winching the parts 1by rope over an imaginary water gap, assembling the gun and firing a shot. The bandaged limbs of the competitors were a sure indication of the wounds inflicted during practice.

It was about this time that I received notification of my posting on return to Australia. The previous December, I had written to the Director of Communications, Colonel Clarke, requesting a regimental command, preferably 5 Signal Regiment in Dundas, NSW. He agreed to my request for a command but advised me that I would be going instead to 6 Signal Regiment, Watsonia, Victoria. He further advised that it was essential for my career that I have a command but this would probably mean I would miss the Joint Services Staff College. I responded that his proposals suited me and I would look forward to 6 Signal Regiment. Soon after I received my Posting Order to the Joint Services Staff College with a letter from the Military Secretary stating that this was essential for my career and that I was now too senior to command. I state the facts without further comment!

* * *

In August I took my family on a two week holiday to the tiny Greek island of Alonissos located in the Aegean Sea and a member of the Sporades group of islands. At the time the island was just emerging as a tourist destination and was relatively unspoilt. We stayed in a resort and spent most of our time swimming and generally relaxing.

I went to Mass on our first Sunday there and caused somewhat of a stir. Not having been in a Greek Orthodox church before I entered the main door, genuflected and took up a standing position at the back. Everything stopped, the priest glared at me and most of the congregation turned and starred. I had inadvertently stood on the female side but was soon nudged onto the appropriate side and Mass continued.

The return trip to the UK was horrific. We travelled by ferry to Athens as planned and arrived at the airport in good time for our evening return flight. Soon after checking in we were informed that the London air traffic controllers were on strike (again) and that there would be a delay of some hours. The airport departure area soon became crowded with stranded passengers and we were soon herded out of the terminal by police apparently concerned for our well-being in the un-airconditioned space. Whatever the motive it was really quite confronting to see the vigour with which passengers were moved.

This was not the only time I had been inconvenienced by strikes at Heathrow, whether by air traffic controllers or baggage handlers. The worst experience I had was on a trip to Germany when I had moved through immigration to the Departure Lounge just minutes before the air traffic controllers went on strike. All passengers in the lounge went to Limbo as were not permitted to exit and had to remain where we were until the stoppage ended some twelve or so hours later.

* * *

The last of my escort tasks involved Colonel Ian Meibusch, mentioned earlier, from Materiel Branch and the General Manager of the Defence Communications Division, Mr Ian Maggs.

I cannot remember the precise reason for the Meibusch visit but it involved us visiting Munich in late September. There was no hotel accommodation in Munich due to Oktoberfest but Siemens was able to secure us rooms in the German Army Signal School Officers Mess. Our stay in the Mess was quite amusing. Unusually, there were not many English speakers in residence but we found common ground in the bar. Ian and I taught the German officers the 3 Battalion regimental song, 'We're a Pack of Bastards' which the Germans thought was one of the greatest of all marching songs and many steins were broken keeping time as we sang. Another highlight of this particular trip was provided by Siemens when they arranged for both Ian and me to conduct one of the Beer Hall bands!

Ian Maggs, whom I had known for some years, had an extensive visit programme in support of his DISCON Project which included visits to the Ministry of Defence, the Communications-Electronics Security Group and a number of the major communications firms in the UK plus a quick trip to Europe to firms in Amsterdam and Munich. What was interesting about this trip was the level of effort made by industry to ensure that Ian had a smooth trip and saw all that was best.

I was particularly impressed when Plessey put its executive helicopter at his disposal while in the UK.

As you would imagine, I spent many hours discussing the DISCON with Ian. These discussions left me feeling concerned about the cost/benefit of the Project, particularly as much of the cost was driven by the provision of high quality secure speech.

I mentioned earlier in this story how I had become interested in Vocoders as a low data rate alternative to high data rate secure speech. Vocoders come in two basic technologies, either Channel or Linear Predictive systems. I do not propose to delve deeper into these alternatives except to say that the output is a 2.4 Kbps data stream which can then be encrypted and passed over a standard 3 kHz channel. The equipment costs are high but the bandwidth requirements are low.

My interest in Vocoders went back to my MALLARD days when I saw how the requirement for high quality (enabling speaker recognition) secure speech led to the requirement for 32 Kbps data channels. What follows from this is that long range communication links could not only be realistically provided by satellite systems.

I had the opportunity to follow my interest with the Secure Speech Research Group and with various UK and European equipment providers. I was also able to acquire a good understanding of the UK government secure speech system which was vocoder based. As a result of this work and the cost/benefit concerns that had arisen during my discussions with Ian Maggs, I prepared a paper 'An Alternative to DISCON' that I forwarded to my masters back home. I'm afraid it received a chilly reception due to the number of vested interests involved.

* * *

The posting to the UK had been both interesting and rewarding but after travelling to and from work in the dark for two years, I was ready to return home. The children who had been reluctant to leave friends in Australia were now experiencing the same pangs on the prospect of leaving the UK.

My replacement, Lieutenant Colonel Ross Thomas, had arrived and I had a busy time showing him around and making my farewells. Our sea baggage had left and I had delivered the SAAB to the agents to have it shipped home. We spent the last week or so in the Waldorf Hotel in the Strand which also allowed me to indulge myself with a few Gilbert and Sullivan performances.

Just before leaving London, the promotion of Keith Morel to Colonel and his posting as Director of Communications was announced. This move was very popular within the Corps of Signals and I looked forward to catching up with him in Canberra.

As I did not have to report to the Joint Services Staff College until the third week in January, we had decided to take the long way home via the USA.

We spent Christmas with Jim McKeon and his family in New Hampshire. I very much enjoyed meeting up with Jim but our two weeks with him was marred by the fact that his wife, Carol, was angling for a divorce. My own home life had deteriorated somewhat over the preceding two years, so Jim and I spent many hours away from the house with his cronies. Despite the unpleasant atmosphere, we were treated to a White Christmas with all the trappings and enjoyed some winter sport.

The next stop was Los Angeles for a belated trip to Disneyland followed by a few days in San Francisco before heading home.

We returned to our house in Curtin to find it in a filthy state and with considerable damage inside and outside. To make matters worse, the Agents had already returned the bond money to the tenants so we all spent time cleaning up and fixing the garden.

Sea baggage arrived on time and we were able to quickly arrange from items left in store to be returned. The SAAB arrived in Sydney so I made a quick trip to recover it from a North Sydney dealer. The car was undamaged but the rear speakers, of a rather stylish spherical design, had been removed. The thief had simply cut wires to the speakers so there was not too much damage to the car itself. I did run in to some trouble with Customs as an officious agent questioned me closely on the time I had spent in Europe, claiming that I had not used the car for the mandatory two year period to avoid import duties. The matter was finally settled in my favour but not until after a great deal of argument. While waiting for delivery of the SAAB, I had purchased an old VW Wagon as a second car. Having been out of the country for two years, I was very much occupied in completing many administrative tasks with both civilian and military authorities; making courtesy calls on the Director of Communications, Colonel Keith Morel, the Military Secretary and others; presenting briefings to interested parties on my time in Europe; and generally clearing the decks before becoming a student again.

JOINT SERVICES STAFF COLLEGE

Attendance at the Joint Services Staff College, JSSC, was generally required as prerequisite for a Defence Central appointment at lieutenant colonel equivalent level which tended to be a prerequisite for promotion. Obtaining the JSSC qualification was not unlike the Staff College one in that it was no guarantee of promotion to higher rank but it was certainly not a hindrance. The JSSC Course aimed to improve our understanding of how other Services functioned as well as providing a good background for operating at higher levels of command and staff. It also provided an opportunity to establish a circle of contacts, if not friends, that would be helpful in dealing with future Tri-Service issues.

The Commandant was Brigadier John Salmon and we were his first course. He had some administrative support and eight members of the Directing Staff. The College was located in Western Creek and generally known as 'the sewerage farm' due to its proximity to that facility. There was a good library, a central lecture theatre and ample syndicate and study rooms. We had a good Mess, a tennis court and a small golf practice area. The reputation of the College was such that it was an easy matter to attract visiting lecturers to talk to us on a wide range of topics.

The course itself was of six months duration and was run at a reasonable pace which offered some time for reflection and personal research. The syndicate system was very much in vogue and most of our work was done in syndicates of eight with an allocated member of the Directing Staff tasked with assisting our deliberations and providing comments for student assessment. Students were allocated to syndicates to provide a balance of Service representation and each syndicate contained an overseas student plus one of the civilians. The syndicates changed a number of times during the course so we all got to know each other well by the end of the six months.

In 1978 the student body consisted of forty students: seven Navy, fourteen Army, fourteen Air Force and five civilians. Included in this number were six foreign students and one Army Reservist. Except for the senior student, Captain (later Rear Admiral) Oscar Hughes, the military students were all at lieutenant colonel level and many were within one to three years of being considered for promotion. Between us we had had a wide range of experience in command and staff appointments which led to some lively discussions and was a boon to the learning experience. I would also say that most of us were comfortable with our standing within our Service and were both willing and able to challenge the status quo and forcefully argue our own points of view. This was not always welcomed by visiting lecturers, the Commandant and the Directing Staff.

I had never met any of the RAN, RAAF or civilian students before arriving at the College but I was delighted to find among my Army colleagues, one RMC classmate, Lieutenant Colonel (later Major General) Mick Jeffery, one from the RMC Class of '57, Lieutenant Colonel (later Lieutenant General) Don McIver of the Royal New Zealand Artillery, and two from the RMC Class of '59 Lieutenant Colonel (later Major General) Horrie Howard and Lieutenant Colonel Brian Hughes. We five from RMC were treated by Brigadier Salmon as long lost brothers as he had known us all at RMC.

The course commenced on the 23 January 1978 and we were soon set into a routine of central presentations followed by syndicate discussions and syndicate or individual presentations. We were each given a research task, almost a sub-thesis, to complete in parallel to our ordinary course work.

Our first trip away was the Services visit RAAF, Amberley; HMAS *Watson* and then a sea voyage; the School of Military Engineering, Sydney; and finally a firepower demonstration at Puckapunyal near Seymour, Victoria. This was the first of many trips in RAAF C130 aircraft.

The highlight for me was doing a jackstay crossing between the DDG we were on and HMAS *Supply* while underway. This involved being hauled between the two ships in a harness that rode on a steel cable over a gap of about 30 metres. The hauling was done by sailors who very much enjoyed causing their passengers to bob about on the line and getting close enough to the sea to get a little wet. Anyway, I enjoyed the experience!

The firepower demonstration was a cause of embarrassment to both the Army and RAAF students – the former because of a recorded commentary that lost synchronisation with what was actually happening and the latter because of a less than perfect bombing run by an F111.

While on the subject of F111s, we had as a fellow student Wing Commander Gill Moore who had spent a number of postings piloting the beasts. He regaled us on many occasions with stories of flying these aircraft with ground following radar activated and admitted that the only way he could cope was to close his eyes, put his head between his knees and hope the system was working properly.

Some time later in the course when we were discussing the costs of training and education, Gill came up with the statement that he had done some calculations on the back of an envelope and estimated that the total costs of training an F111 pilot including initial training, conversion, trips to the USA, flying time, etc was $6M. Henceforth, he was nicknamed the $6M man in line with the main character of a current popular TV series. At the period following Gill's statement, Mike Webster, an Army Aviation Corps pilot, set us in hysterics by proclaiming that he had done some calculations on the back of a postage stamp and estimated the total costs of training an Army pilot to be $10.50!

We had many visiting lecturers and the general practice was to have them in pairs with one performing before lunch, both having lunch in the Mess with staff and students with the second speaker following after the meal. This usually worked very well as it allowed very worthwhile informal discussions. There was one incident, however, that carried a salutary lesson for presenters. The morning speaker had opened his address with a very funny joke followed by an informative and well delivered lecture. Unfortunately, the afternoon speaker opened with precisely the same joke as his predecessor and didn't even raise a giggle. This virtually destroyed the speaker who did not recover his composure. The moral of this story for speakers is to be very careful when opening lectures with a joke and always be prepared for it to fall flat!

We had a number of sporting opportunities during the year with one of the main events being a golf day at Royal Canberra Golf Club. I had not played since Queenscliff and quickly found that playing golf was not like riding a bike – the skill could readily be forgotten. Anyway, I was playing very badly until reaching one of the Par Three holes. I was teeing off last having had a terrible previous hole. None of my playing partners had hit the green but all were close. I took my stance, duffed the shot which flew close to the ground until bouncing off the head of a grazing kangaroo to land within a foot of the pin. I won the hole but the method of doing it completely fazed my partners and ruined their concentration for the rest of the round. I didn't play golf again during the rest of the course but I'm pleased to say it did not put me entirely off the game and I took it up with a vengeance many years later.

As I said earlier, most of the students were more than willing to air their views and challenge lecturers. The most memorable example of this was during a presentation to Australians only by a senior official of the Defence Intelligence Agency. In question time, a number of students raised the issue of perceived over-classification of some intelligence product. This raised the ire of the Director who then produced a slide, marked SECRET, which purported to be of a secret installation behind the Iron Curtain citing that this material was too sensitive to acknowledge the source. While the Director harangued us, one of the RAAF students, whose name escapes me I'm afraid, went to the library and returned with an Aviation Week magazine with the same illustration and saying *"I rest my case"*.

As you might imagine, the room was in a polite uproar with the DIO official becoming increasingly excited and somewhat over the top in his description of us. To his credit, the Commandant came in very strongly on our side with the result that a few days later our programme was altered to allow the Australians to attend an excellent briefing in the DIO building.

The course was drawing to a close and we were preparing for our overseas trip when I was struck down with a tooth abscess. I started the morning with a dull ache but by lunchtime my face had swollen to the point that I could not see out of one eye. I went off to the dentist but he could do nothing until the swelling went down. As a result I was admitted to hospital and filled with penicillin. The treatment worked, my tooth was removed and I was able to join the trip.

The course was split in two for the overseas trip. We all travelled to Manilla via Darwin in a C130. We had an evening and day together before my half of the group went off to Thailand while the other half toured the Philippines.

On arrival in Bangkok, we were met by Colonel Lachie Thompson who was the Defence Attaché at the time. I had played drums in Lachie's RMC jazz band and we were very good friends. Lachie had had multiple postings to Thailand, was fluent in the language and regularly played clarinet with the King. Lachie's relationship with the King meant that he had an unprecedented level of access into the Thai defence hierarchy and was able to organise a comprehensive set of briefings and visits.

Our C130 returned to the Philippines to be at the disposal of the other half of the course while we had a RAAF DC3 that was in beautiful condition to ferry us around Thailand. We had one incident in the DC3 when the junior pilot had been given the task of landing the aircraft on one of our shorter journeys. We were on finals and things did not seem to be going to plan. The pilots on the course started to look a little worried which was quickly noted by the rest of us. I must say we were taking it all in our stride when suddenly the pilots on the course went white knuckled and braced for a crash. We did, in fact, land safely without damage to the aircraft. What followed was of great interest to the brown jobs at least. With the senior pilots on the course giving the DC3 captain considerable policy guidance.

We spent five days in Thailand and made visits to various Royal Thai Army and Navy establishments in Nong Khai, Chang Mai, Phitsanulok and Pattaya. In Phitsanulok, we were entertained at a concert given by the local commander. Lachie had been asked to play his clarinet for the assembly. He announced from the stage that he would play a recent composition of the King. The audience went mad and it was easy to see why Lachie was so well regarded.

There were, in fact, two visit programmes – the official one which included the whole student body and a second out-of-hours one which Lachie gave to a few of us that he knew well. The stand out of the latter, was a visit to a particular conference room in one of the Royal Thai Army headquarters buildings where the coups took place. Lachie explained that the senior Generals would sit around the table and go through an exercise not unlike bidding in Bridge. Each would open with a listing of the major units loyal to them. If this did not produce a result a second round would follow with listing of minor units or perhaps would challenge claims made in the opening bids. Eventually this process would lead to a clear idea of where the power laid and the winning General would then advise the King and the coup would be completed without loss of blood. Of course, I cannot vouch for the accuracy of Lachie's description but it seemed too good a story to omit from this tale.

It was not all work on the trip and we had ample opportunity to see the local sites and enjoy the local food. I particularly enjoyed the silver market in Chang Mai where local craftsmen created the most delicate of delicate silver wire filigree jewellery, the source of silver being old coins from Burma.

We finished our Thailand tour at the resort city of Pattaya. We had visited a Royal Thai Navy base during the day and had noticed a large number of vessels in various stages of repair ranging from the operationally ready to rusting hulks. We were informed that the number of Royal Thai Admirals was determined by the number of ships on the inventory which certainly explained what we had observed. That night in Pattaya, Lachie used his influence with the proprietor of our hotel for we two to displace the hotel band and play some jazz, much to my delight.

We had been joined by the other half of the course in Bangkok and started the return journey via Bali in our trusty C130. The C130 did not prove as trusty as we had imagined and we were 'forced' to spend an additional day and night at a magnificent tourist hotel.

We finally arrived back in Canberra for a rest day and then into the final two weeks of presentations, submission of projects, farewell drinks with the staff, addresses by the Chief of the Defence Force Staff and the Secretary, etc before facing the Commandant for our assessment. The interview with the Brigadier was painless – I cannot claim that I shone on the course but then I was nowhere near the bottom. The Brigadier gave me a frank assessment but then apologised because he felt that he could not, after consultation with the Directing Staff, recommend that I have a future posting on the said Directing Staff. I managed to retain a stoic expression and accepted this advice in a suitable chastened manner while inwardly rejoicing.

I must say that I enjoyed the course which I found stimulating and I certainly valued the opportunity to meet with representatives of the other Services. A number of the students were stand outs and it would have been impossible to place a bet against the future success of Oscar Hughes, Horrie Howard, Mike Jeffery and Don McIver. I'm sure that other members of the class did very well in their careers but I lost touch with many of them so I cannot readily confirm this.

THE JOINT STAFF

I had been posted to the Joint Communications Branch (JCB) of the Joint Staff, initially in an administrative position that was changed to the Staff Officer (Operations) following the intervention of my old friend Brigadier David McMillen who was the Director General (DGJC). I was absolutely delighted to be on David's staff as I had worked under him as a Lieutenant, Captain, Major and now Lieutenant Colonel. David was a hard taskmaster but he firmly believed in allowing subordinates to get on with their jobs while always being ready to advise and guide. I never heard him raise his voice to a subordinate – if a 'rocket' was in order, he would invite the miscreant into his office and simply say:

"I'm disappointed in you – March Out."

It happened to me on more than one occasion and I always left his office feeling much worse that I would have felt if I had been subjected to a harangue.

By the time I arrived in the JCB, implementation of the Tange Report had been in progress for some five years. There was still a fair amount of suspicion directed at the growth of influence of the civilian elements of the Department. Until the implementation of the Report in 1973, the Department of Defence was of relatively little consequence in the Commonwealth government; each Service (Navy, Army and Air Force) had its own separate department with its own minister. There was a clear perception that the civilians in the Service departments were part of the team and very supportive of the aims and aspirations of their Service. There was certainly competition between the individual Services and Departments over budgetary and other matters. This competition did not go away under the new arrangements. In fact, I would say that there was a deliberate policy within the now powerful Department of Defence to foster a confrontational approach by the Services, particularly with regard to equipment acquisition.

The diarchy, Secretary of Defence and Chief of Defence Force Staff, to advise government and administer the Australian Defence Organisation was well established and, in my opinion, worked reasonably well. There were, of course, tensions and these flowed down through the hierarchy. One of the contentious issues in my time was the establishment by Sir Arthur of a table of 'equivalence' between military ranks and civilian classifications. This was used as a means of structuring the many organisational elements within the Department that contained military and civilian members and was the prime source of classification creep.

Another thing that did strike me at the time was the nonsensical attitude that seemed to prevail among some senior public servants that academic qualifications held by military officers were somehow less acceptable than those held by civilians. It is well recorded that Sir Arthur felt that senior military officers of his generation lacked the broader humanistic education that would allow them to make other than technical contributions to wider defence policy. In line with this, Sir Arthur was instrumental in the decision to set up a primary tri-service college, to be known as the Australian Defence Force Academy.

Without doubt a posting to the Joint Staff was the pick of the bunch of postings to Defence Central, known to many as the great grey sponge. I would not like to suggest that the only worthwhile military appointments in Defence Central were on the Joint Staff, but there were certainly many where uniformed members were included in largely civilian staffed sections to give credence to the policy of departmental integration. In many of these, the incumbents lacked real responsibilities and challenges and constantly sought to be reposted.

The Joint Staff was led by the Chief of Joint Operations and Plans (CJOP) who, at the time, was Vice Admiral Peter Doyle. I did not have regular contact with the Admiral on other than the Friday morning briefings except for one memorable occasion which occurred in my second year in the appointment – more on that later.

The JCB was an integral part of the Joint Staff and performed many functions at the national and international level. The DGJC was the senior member of the ADF dealing with communications matters and was the primary source of advice to both the CDFS and Secretary. He was also the Australian Principal on the Combined Communications

Electronics Board (CCEB) which was formed during World War II to ensure, not always successfully, communications interoperability between the Allies. The role of the CCEB had hardly changed since the war and many of our national responsibilities flowed directly from it.

The national responsibilities of the JCB were pursued through the Defence Communications Committee which was chaired by the DG and comprised the single Service senior communicators. In 1978, these were the Director Naval Communications, Captain (later Commodore) Harry Adams; Director of Communications-Army, Colonel (later Brigadier) Keith Morel; and the Director Communications Electronics-Air Force, Group Captain (later Air Commodore) Reg Rowell. There was a subordinate committee, the Defence Communications Coordinating Committee, chaired by my section head Captain Richard Arundel, RAN, which included myself, the RAN representative, Commander Doyne Hunt and the RAAF representative, Wing Commander Rex Bean. The operations side of the Branch did not operate in a vacuum and many of the tasks to be performed contained an engineering aspect and so I worked closely with Wing Commander Dave Millar. Other colleagues in the Branch were Wally Rothwell, RAN; and John Gordon and Dick Carfax-Foster, both of whom were members of RASigs.

* * *

I had no sooner arrived in the JCB when the Telecom dispute broke out. Telecom had announced it intended to computerise telephone exchanges and carry out all maintenance through a number of Exchange Maintenance Centres which would remove the need for a maintenance crew at each exchange. The technicians union (ATEA) then refused to carry out repairs and maintenance and particularly targeted Telecom's centralised revenue recording equipment. The union's ability to target sensitive telecommunications facilities meant it could threaten business as well as government. The fact that the union was fighting over the consequences of technological change at a time when much of the public had similar concerns gained the union a considerable level of sympathy.

In mid-August, Telecom started standing down technicians which had little effect but the bans were proving very effective. By 21 August, telecommunications in New South Wales and Western Australia were near to collapse and just a few days later breakdowns had spread to South Australia and Victoria.

The JCB became involved almost from the outset as most government communications were provided by Telecom either over its network or by private networks using cables leased from Telecom. Defence communications were highly reliant on Telecom but we did have a means of back-up through the HF Radio networks.

I think my immediate superior, Captain Richard Arundel, must have been on leave because I became the point of contact for Defence dealings with government. I was involved in daily briefings with government and had a few moments of fame when I was referred to as a Very Senior Officer from Defence. This title much amused David McMillen who proclaimed that if I was a VSO he was a VFSO. I will leave it to the reader to unravel this abbreviation!

We were looking at a complete collapse of the Telecom system as exchange after exchange was taken off air and the union refused to carry out repairs. I would not suggest there was outright sabotage but there was a higher than usual interruption to the Sydney-Melbourne Cable due to accidental cable cuts and it was thought that many exchanges, all of which were housed in climate controlled buildings, closed down because of air conditioners being accidentally switched off. A new term, 'educated elbow' appeared and was cited as a probable cause for the marked decrease in the reliability of telephone exchanges. Union solidarity was growing with telegraphists in the Sydney GPO stopping work and Adelaide telephonists threatening to walk out if asked to do technicians' jobs which meant that even simple tasks like resetting a circuit-breaker that would restore a service were banned by the unions.

There had been some discussion about using military personnel to man the exchanges to restore service but this did not eventuate. The closest we got was the Army deploying a communications detachment to the PM's property at Nareen to keep him connected to the seat of government.

One of the early challenges for Defence was the loss of the Telecom line from one of the intercept stations to the Defence Signals Directorate (DSD) which was still located in Melbourne. Provision of some sort of substitute was within my purview so I made contact with the appropriate DSD officer seeking such details as acceptable error rate and capacity. I ran into a brick wall and was told that I did not have the required level of security clearance. My response was that without the information I could not supply the required communications. When I informed the DG he became quite apoplectic and proceeded to make a few phone calls. The upshot was that I was invited to Melbourne for a few days to be briefed on every conceivable aspect of DSD operations and to meet the principle members of staff which was of great assistance later on.

The dispute was finally resolved after the President of the ACTU, Bob Hawke, brokered a deal with the Arbitration Commission. As a further result of arbitration, the ATEA claimed that they had regained control over the introduction of new technology but this claim was tested a number of times in the following years. Once the dispute was settled, telecommunications services were restored with remarkable speed giving credence to the claims of 'educated elbow'.

The Telecom dispute had long term consequences for communications for Defence as it highlighted our vulnerability to union interference. This dispute was quoted for years as a primary driver for development of a communications system which would be entirely independent of Telecom. As the years went by a more pragmatic approach was taken and some grandiose plans for a Defence owned, nationwide network of broadband microwave channels were scaled back.

* * *

With the end of the Telecom dispute, work in the JCB got back to normal. I spent much of my time working with the Services on contingency plans that were being developed for offshore deployments that might be required to support friendly governments or recover Australian nationals from countries in our region if the local situation was a cause of concern. I was also very much involved with contingency plans to cover any further disruption to government communications through industrial disputes or natural disasters. I had established good relationships with a number of sections within DSD during the dispute and spent a fair amount of time becoming familiar with its various operations and methods of work.

It was not all hard work and many of us in the Joint Staff gathered socially from time to time building on relationships forged at the JSSC and generally improving our understanding of our various functions and the idiosyncrasies of the three Services.

One of the vehicles for social gatherings was the ISDG formed towards the latter end of 1978. ISDG was translated for the benefit of our superiors as the Inter-Service Discussion Group with an alternative being the Inter-Service Drinking Group. We used to meet every two weeks or so in the back bar of the Ainslie Hotel. We generally restricted our meetings to the lunch hour but, if we did run overtime, we found that explaining to our section heads that we had been attending a meeting of the ISDG was readily accepted. To add credibility to the ISDG, we had, in the best traditions of the Services, commissioned the manufacture of a tie. After much discussion, we decided to use the existing Joint Staff tie, grey with red, light blue and dark blue stripes, with the initials ISDG embroidered in the centre. The group and the tie were still in use when I returned to the Branch in 1985.

* * *

My married life had been in decline for some years and in an attempt to resurrect it, Yvonne and I attended a number of sessions with a marriage councillor. These sessions were not that helpful and we decided to have a 'trial' separation. As a result I moved into Brassey House in early October 1978. I was not enthusiastic about this move but there did not seem much prospect of our marriage lasting.

Brassey House was at the top rung of a hierarchy of government owned hostels and was home to many public servants of various levels. It was used by Defence to house single or married unaccompanied officers on posting to Canberra or attending the JSSC. There was always a small itinerant population of military and civilian visitors to Canberra as the place was comfortable and well run. I had visited the place once or twice while at JSSC and thought it would suit, at least in the short term.

I have to say that after eighteen years of marriage I was not looking forward to spending the rest of my life as a bachelor. I had not given up hope of a reconciliation but I thought this unlikely. I expected that Yvonne would eventually ask for a divorce which would leave me unable to remarry in the Catholic Church unless the marriage was annulled. I thought that this would be unlikely and did not think more of it at the time.

Brassey House had its fair share of characters not the least being the Manager who was a retired Major who had served in the Royal Artillery. He ran the place as much as he could along the lines of an officers mess and generally did it well. He did have quite a temper and I felt sorry for his wife who was often the butt of it.

Many of the long-term residents were quite institutionalised and carefully nurtured real or imagined hurts, disputes and grudges that were often amusing but at times rather pathetic.

One lady had brought a chair from her previous hostel which she placed in the upstairs TV Room in her favourite viewing spot. This was her chair and she apprised all new residents of this fact. There was a male resident who would regularly sit in her chair to goad the lady into a temper tantrum. He, himself, was quite odd and every two weeks or so he would load all his bed linen into one of the toilets which would drive the Manager into near apoplexy, particularly as the Manager seemed powerless to punish this behaviour.

There was another lady who had once been Secretary to the Chief of the General Staff which gave her, in her mind, such status that she would only allow officers whose rank was major or above to sit at her table.

There was another man there who I never saw speaking in all the time I was in residence. He would appear for breakfast at the first bell, eat a huge meal without as much as a glance around him and immediately depart for his office (I never did find out where he worked or his classification). On the very odd occasion that I was there for lunch, I saw him enter the dining room, eat his fill in silence and again depart immediately to return to his work. In the evenings he would do likewise but would go straight to his room.

There have been many stories told about Brassey House but most are apocryphal. I stayed there until March 1981 until one fateful Saturday morning when I realised that I was becoming as institutionalised as the other long term residents and that the room I was living in was smaller than the one I had had at RMC. I saw an Estate Agent that day and had moved into a flat in Holder by the middle of the following week.

* * *

The year came to a close with Brigadier David McMillen leaving the Branch to return to an Army Headquarters appointment. His replacement was to be Air Commodore Reg Rowell who had been promoted into the position from Director Communications-Electronics Air Force and was set to join the Branch in early 1979.

I was extremely sorry when David left the Branch as, apart from being an excellent boss and good friend, he helped me through the difficult time of coping with my separation.

* * *

I started 1979 with a week of Duty Officer. This could be as dull as dish water but occasionally it could become quite hectic, particularly if there was an ongoing natural disaster and calls were being made by State governments for assistance. Trying to organise assistance with the individual Services was not always easy and there were occasional hiccups. These hiccups invariably led to 'you said – we said' recriminations which could become quite nasty. I regret to say that we had eventually to resort to acquiring voice-activated recording equipment to resolve such issues.

During this week I also had a call to report to the Military Secretary's office to have my photograph taken as a prelude to my interview with the Promotion and Selection Committee – but more on this later.

Early in the year I became involved again with satellite communications by becoming a member of the National Satellite Working Group. This work was always interesting and rewarding as it was contributing to the enhancement of national infrastructure.

Not long after taking up the appointment of DGJC, the Air Commodore arranged a visit to the Joint Facilities at North West Cape. The operational and engineering aspects were of interest so David Millar and I joined his party. This was the first of what was to be many visits to the Joint Facilities.

There was an inauspicious start to the trip. We arrived at RAAF Fairbairn to catch a C130 to Richmond, where we were to join a USAF C141 making its regular replenishment run to the Cape, only to find that the regular Monday flight had been cancelled. So it was off to Richmond by road in my ancient VW Wagon. It was not a very inspiring trip as the new DG hardly spoke a word throughout the journey.

We flew from Richmond directly to North West Cape tracking right over the centre of the country. I spent most of the trip on the flight deck enthralled by the grandeur of the GAFA. I had never heard the term before but I was subsequently informed it meant the Great Australian F*** All.

The C141 was a huge aircraft capable of carrying 154 troops or some 32, 000 Kg over short distances. It had high mounted drooping wings which lifted when airborne. I was surprised to find that the C141 had a crew of only five: pilot, copilot, engineer, navigator and loadmaster. The aircraft was highly automated and it seemed to me that the only real workload was that of the engineer who kept an eye on fuel consumption.

The station had been commissioned in 1967 as the US Naval Communications Station North West Cape which provided HF and Very Low Frequency transmissions to ships and submarines. The station was renamed US Naval Communications Station Harold E Holt in September 1968 in memory of the former prime minister who disappeared while swimming some three months after the station was commissioned.

The initial lease did not allow Australia any degree of control over the station or its use. After the election of the Labor government in 1972, Defence Minister Lance Bernard commenced negotiations on the operation of US bases in Australia. In 1974 an agreement was reached which allowed for a RAN officer to be the Deputy Commander and assigned technical and maintenance roles to Australian personnel. The base was controversial as its main use was to relay orders to nuclear armed submarines.

The base was fascinating from a technical point of view as the main VLF transmitter had a power of 1 Megawatt. Because of the low frequency, there needed to be a minimum of metal in the construction (the frames for the output transformers were wood) and special arrangements were required in the building construction as heat generated in the concrete floor could cause rapid degeneration. VLF also meant that metal objects could not be carried while in the building. Another interesting technical issue was the need for superfine filters on the air intake for the HF transmitters. In the early days of the station, the transmitters were failing regularly due to overheating caused by filters becoming clogged.

The station features thirteen tall radio towers. The tallest tower is called *Tower Zero* and is 387 m (1,270 ft) tall, and was for many years the tallest man-made structure in the Southern Hemisphere. Six towers, each 304 metres tall, are evenly placed in a hexagon around Tower Zero. The other six towers, which are each 364 metres tall, are evenly placed in a larger hexagon around Tower Zero. We were all invited to ride the elevator to the top of Tower Zero and those who did received a certificate to prove it. I opted to stay on the ground and by the look on the DG's face when he got back to earth confirmed for me that I had made the right decision. After an overnight stay, we returned to RAAF Pearce by charter flight and then to Richmond by C130.

* * *

We were no sooner back from North West Cape when I was advised that my Promotion and Selection interview was to be held on St Valentine's Day. I wasn't sure if this was some sort of omen of success but thought it could not do any harm.

There was an initial interview in the Military Secretary's office in the morning to advise candidates of the procedure and to name the members of the Selection Board. My actual interview was to be in the Chief of Personnel's office at 15:00. There was a long delay between the candidate before me leaving the room and the appearance of the Military Secretary, Brigadier Gordon Fitzgerald, to invite me in. As I approached the Board I had just noticed that the interviewee's chair had been replaced by a barstool when Gordon remarked *"The Board waited to see you clearly so we thought a stool would be best"*. While I climbed onto the stool the Chief welcomed me to the interview and then asked the other Board members if they had any questions. None did so the Chief thanked me for coming and bid me good-bye. The whole process took about a minute so I felt that I was either home and hosed or I would never ever see a red hat. There was no follow-up, so it was now just a matter of waiting.

* * *

I had become concerned about the amount of time I was spending in the bar at Brassey House, so in early March I enrolled at the Canberra College of Advanced Education to do a Diploma of Advance

Mathematics. I had no particular need to take the course but it seemed like a good idea to keep me occupied after hours and to receive some mental stimulation. I persevered with the course for most of the year but I was having difficulty completing assignments on the CCAE computer as it was assumed that all students would have some back ground in computer operation and programming. I had neither of these skills but managed to survive until the latter half of the year when the set problems became more complex. I was struggling at the student terminal one evening when the screen flashed at me with the message: 'If you persist in giving illegal commands you will be suspended'. This was just too much – I had received 'rockets' from my seniors on many occasions but to get one from an inanimate object tipped me over the edge and I withdrew from the course.

* * *

I had missed out on operational service in Vietnam but a second opportunity arose in March 1979. Australia had agreed to provide a contingent of Royal Australian Engineers to assist in nation building projects in Namibia. I was nominated to conduct a communications reconnaissance before any troops were deployed. I was subjected to a round of extensive briefings, given a number of exotic inoculations and sent to the Q Store to be properly equipped for the task. I was issued with some warm weather clothing, a steel helmet which I was to have painted UN Blue on arrival and various other items. Among these items was a length of thin nylon cord. I queried this and was told by the Quarter Master Sergeant that this was to attach my clasp knife to my belt. I informed the QMS that I did not have a clasp knife and asked for one. He replied there were none in store but insisted I take the cord.

Preparations went on apace and I was placed on 24 hours notice to move. I had prepared all my kit and had signed for my One Time Letter Pads and a series of Credit Cards and was ready to go. To ensure that I could meet the level of notice I was on, I was issued one of the new fangled Beepers – all very exciting.

Not long after being issued with my Beeper, I was attending a party when it went off. Being a fan of Get Smart, I immediately whipped off a shoe and spoke into the heel before moving out of the room to check the message. The message turned out to be only a check but when I returned to the party room I overheard one of the female guests asking in an awed voice "*Is he a spy?*" which was a source of great amusement to my hosts and other guests.

I remained on 24 hours notice for about two weeks and this was then extended to 48 hours and this level was maintained for a few months until the period of notice became two weeks. While I was waiting for my call, the engineers were frantically painting their vehicles and plant white in anticipation of their deployment. The engineers finally did go to Namibia but I got posted at the end of the year and so missed my chance of a medal.

* * *

Apart from playing hockey on Winter Saturday afternoons, I was pretty much at a loose end for most weekends until one Saturday in late February, I was invited to join a group from Brassey going to an afternoon session of the ACT Jazz Club at the Dickson Hotel – the 'Dicko'. The visit rekindled my interest in jazz and I became a regular. The usual band was the Black Mountain Band which played traditional Dixieland jazz. It had the usual line up of trumpet, trombone, clarinet, piano, bass and drums and was led by Terry Thomas. After attending for a few weeks I approached Terry and told him of my background with Lachie Thompson and from then was invited to 'sit in' from time to time. This was a great source of enjoyment and I often visited a number of jazz venues over the weekends. I made many new friend at the Club including Shirley who became my wife a few years later.

The Jazz Club filled a social void, as I soon experienced what most people do when they become divorced. Many friends did not want to be seen as taking sides and some husbands saw single males as some sort of threat, so invitations to parties dwindled quite rapidly.

I found the first twelve months of separation extremely difficult. I had met a number of females but was reluctant to enter any serious relationships as I continued to harbour a desire for reconciliation.

* * *

Air Commodore Rowell was a very different man to David McMillen and I, and many of my colleagues, found it difficult to relate to him. He was very good at his job but he tended to remain in his office and not wander among his staff. He went out for lunch each day and never advised us where he was going. This, of course, was his prerogative but it did occasionally present the Branch with problems. I mention this, not as a criticism of him, but to provide some background to one of my more memorable experiences.

The Branch was involved from time to time in classified projects that were known only to the DG, Captain Arundel and myself. One of these involved stashing some funds in an obscure account which we confidently thought would be undetected. How wrong we were! The CJOP, Admiral Doyle, had somehow discovered the money and demanded an immediate briefing. The DG was at lunch at some unknown location and Captain Arundel was on leave so I was at the bottom of the chain. On my way to the great man's office, I rehearsed my lines in preparation for the confrontation. CJOP was not known for his sense of humour and he could be very brusque indeed. When I arrived at the office, he shouted at me to provide an explanation. I responded that he was not cleared to know. As he descended back to earth I quickly added that I could provide him with a detailed brief if necessary. After a quick outline he was somewhat mollified and I was able to return to my office relatively unscathed. In this I was very fortunate as a number of officers felt his wrath with often very deleterious affects on their careers.

* * *

The year wore on and I continued to be very happy in my appointment. The work was interesting and varied and involved a reasonable amount of travel. The second half of the year was broken up nicely by a one month long Defence Systems Management Course run by the Chief Defence Scientist's office. I have only vague memories of the course which is a pity as I have what purports to be a Course Certificate with FAIL in red letters. I seem to recall that I challenged the Chief Defence Scientist a little too vigorously which resulted in my classmates producing the certificate.

Towards the end of the year I was asked by Dr John Alwyn, a senior medical officer at the Russell Offices Medical Centre, if I would help him out by taking on the job of Secretary/Treasurer of the St John Ambulance Association. What was to be a short term involvement turned into thirty years in just about every appointment there was in the ACT.

Major General Ron Grey had become the Chief of Operations-Army and I was becoming increasingly involved with Mike Jeffery on some proposed new functions for the Special Air Service which I must necessarily gloss over. I did have the occasional embarrassment with the General as he sometimes would call me for some Signals advice. I would always respond with the fact that he had a very competent Director of Communications in Colonel Keith Morel. The General agreed with that but he usually insisted on asking *"but what do you think?"*. I would give my advice and would then hot foot it over to Keith Morel to report the conversation. Happily our advice always coincided.

My relationship with Shirley was developing and I had met her three daughters Linley, Deidre and Elizabeth. Shirley had been widowed some two years before we met and the girls were not quite sure how to take me at first but a good relationship did develop over time.

Deidre worked in Defence at Campbell Park. She had heard me talk about sports cars and one day called to advise me there was a note on her Notice Board advertising a MG B for sale. After a quick look I bought the car with great visions of working on the mechanicals myself to keep busy at Brassey. I bought a MG B Handbook and a basic set of tools as a first step.

My first project was to reattach the stop light pressure switch to the engine bay bulkhead. In attempting this, I managed to break a brake-line and had to have the car towed to a repairer. I learnt my lesson and decided that, in future, I would limit my restoration work to washing and polishing the car.

The Corps Dinner was to be held in mid November and I had asked Shirley to accompany me. A week before the dinner, I received a call from the Military Secretary who advised me that I was to be promoted to Colonel on the 10 December, precisely twenty-one years after I had graduated from RMC, and was to be posted as Director of Communications. I was sworn to secrecy and told that the promotion and posting would be published before the dinner.

I was, of course, elated with the prospect and scuttled back to my office in Joint Communications to give thought to the future. I had no sooner got back when John Gordon came in and asked if we could swap duty officer that I was to do the following week until a date in January. I was obliged to accept the proposal to keep faith with the Military Secretary but I must say I felt bad about it.

The Corps Dinner was memorable indeed. Colonel John Kemp, who was Director of Engineers at the time, had been invited to propose the toast to the Corps as tradition demanded. He gave an extremely amusing address comparing the outgoing Director, Keith Morel, and me. It is a matter of great regret that I was unable to get a copy of his address to add to my memorabilia.

A few days later, my RMC Class had our 21st Anniversary of graduation reunion. It was a grand affair starting with dinner on the preceding Saturday, a memorial church service on Sunday and the Graduation Parade on the following Tuesday. We still follow the same format to this day.

And so ended 1979.

DIRECTOR OF COMMUNICATIONS – ARMY

The Directorate had gone through a number of iterations in the years preceding my appointment but had finally settled down to three sections, Operations; Engineering and Technical Support; and Personnel Management, headed respectively by Lieutenant Colonels Mike Collins, Jim Messini and Herbie O'Flynn. D Comms-A was a large directorate with both staff and Head of Corps responsibilities. Apart from myself and the three section heads, there were twenty-five on the posted strength made up of eight majors, two captains, eight warrant officers, two sergeants, two signalmen and three civilian clerks. I already knew most of the staff and I was looking forward to working with them.

I took over as Corps Director in a time of change. Control of the physical assets of the Army fixed communications network had passed to the Defence Communications Division and there was ongoing argument within the Corps on whether we should extract ourselves completely from strategic communications and concentrate on tactical systems. Much of my background had been in fixed communications and I was anxious that the Corps should remain involved as the technical skills were readily transferable between the two and the existence of postings in the fixed network enabled rotation between fixed and field in similar fashion to the RAN's shore and sea postings.

There was also considerable discussion about who should be responsible for computing within the Army. The Corps had been exposed to computer-based message switching through STRAD and other smaller tactical systems over a number of years and many units had acquired PCs to perform some administrative functions. The Ordnance Corps had been using large computers within the supply system and many members of the Army thought that this expertise qualified Ordnance to take on general responsibility for computing. The concept of informations systems was in its infancy and I was determined to keep the Corps involved.

I felt strongly that it was essential for the Director to see and be seen by all members of the Corps and I was determined to travel as extensively as the travel budget would permit. I aimed to visit each regular and CMF unit at least once per year and to be in attendance at most of the important courses held at the School of Signals. I was also eager to maintain a high profile for the Corps so determined to regularly visit the functional commands and military districts.

I felt great pride in becoming the professional head of the Corps of Signals and commenced my new appointment with great enthusiasm tinged with some trepidation that I would be up to the challenge. I was well aware that there were many other members of the Corps as capable and as qualified as I was to do the job but, having been given the appointment, it was up to me to put my mark on the Corps. In this task I was greatly encouraged by the sheer volume of messages offering congratulations on my appointment and wishing me well in it.

* * *

One of my first acts on joining the Directorate was to ask, in my sternest voice, who was responsible for the two cartoons that had appeared in the November 1979 Corps Newsletters. One of these showed a small building in front of a larger one with a sign reading:

Russell Offices

Mini J Block

DCOMMS-A

SO1 PERS MNGT

SO2 OR MNGT

The other showed an open door with a nameplate PJA Evans and another nameplate KP Morelon the floor with a tradesman sawing a few inches off the legs of a chair an a desk with a piece of verse reading:

AROUND THE COMMS DIRECTORATE

THERE IS A LOT OF A'DO

ABOUT THE HEIGHT OF FUTURE STAFF

TALLER THAN 5 FOOT 2.

There was a rather subdued response from Herbie O'Flynn asking why I wanted to know. When I replied that I wanted the originals for my collection there was a noticeable lightening of mood!

Once my reaction to the cartoons became known, there seemed to be a reasonably regular flow of new ones all referring to my stature.

* * *

My first official visit as Director was to my old stamping ground at Dundas where I had been invited to visit to present a number of National Medals. I had been attached to 402 Signal Regiment, renamed 5 Signal Regiment in 1965, while waiting to attend University and had been posted back there in the early 60's. As a result of rationalization studies, the Regiment evolved in 1979 into 134 Signal Squadron with a strength of some 140.

At the due time, I arrived on the parade ground (really a car park) to receive a salute. After inspecting a small guard, I was invited to make the medal presentations. The first recipient was well over six feet tall and I would have needed to stand on the tips of my toes to reach his breast which is hardly dignified. Suddenly a box appeared which provided sufficient elevation to pin on the medal.

A few weeks later, on my first trip to Watsonia as Director, I arranged with the Corps RSM to visit the Sergeants Mess to meet as many as possible of the Corps seniors.

I had barely got in the door when I thought I recognised a familiar face. I said *"we have served together before"* and received the response *"No sir, that was my father"*. Suitably chastened, I continued the rounds with the RSM at my side and a steward one pace behind bearing a silver tray and a decanter of Scotch. After a few minutes I was introduced to a smallish group of technicians and after some small talk, one of them (I very much regret that I did not get his name) asked *"Why don't we have a Banner? RAEME have one and so why shouldn't we"*. I agreed and as soon as I returned to Canberra I put in a request to the Adjutant General seeking approval for an approach to be made to the Colonel-in-Chief, Princess Anne, for a Banner. This was approved the same day and the wheels were set in motion.

I learnt some little time after this visit that some wag anointed me with a new nickname 'The Bionic Budgie' within minutes of my arrival in the Mess.

To complete the trilogy of 'short' stories, soon after the Sergeants' Mess visit, I was the principal guest at the Young Officers' Course Dinner. After the Loyal Toast, the CO 6 Signal Regiment, Lieutenant Colonel Ron Eather, and I were summoned before the Subalterns' Court to see if we were 'up to our jobs'. I have a series of photographs showing Ron and I standing on the head table being charged, solemnly measured with a tape and then fulfilling sentence by drinking a 'glumpha' or two. It was a memorable night.

* * *

You may recall that I bought an old MG sports car towards the end of 1979 and my short career as a restorer. I was, however, still keen to own a presentable vehicle, so I arranged to have the body stripped and re-sprayed and the seats and interior professionally reworked. I did end up with a very attractive car and joined the MG Car Club. I mention this only because later in the year, a rumour circulated within Corps circles that, in order to get promoted, one had to be under 5'6", a Catholic and a member of the MG Car Club!

* * *

Within a few months of becoming Director, I became involved in the Promotion and Selection Advisory Committee. This work along with all matters dealing with personnel was demanding and, at times, emotionally draining. I had already been exposed to the demands of handling promotion and posting issues in a previous posting in the Directorate but I was now involved even more deeply in making decisions or tendering advice that could have a profound affect on people's careers and on their family life.

Making recommendations for promotion to Lieutenant Colonel were particularly difficult as there were always more contenders than vacancies. Recommendations for promotion were based on a close examination of historical performance as recorded in the Annual Confidential Reports. These were not always cut and dried as some people, for reasons outside their control, had a limited reporting history while others had suffered from personality clashes with reporting officers. As one would expect, there were always standouts at either end of the panels being considered and these could usually be quickly settled although not without much soul searching to ensure that there was total objectivity when making comparisons.

The group in the middle was a different manner and placing them in an order of preference was usually very difficult indeed. I'm sure my predecessors and successors all had the same difficulty and I expect all had the same misgivings. However, the work had to be done and done in the best interests of the Army, the Corps and the individual in that order.

One advantage of becoming a Corps Director was the possibility of having previous decisions made by the Promotion and Selection Committee reviewed so that passed-over officers could be reconsidered for promotion. This was not an easy task and required considerable evidence to support a case. I had one success in this area and saw a very competent officer who had been unfairly treated afforded some justice.

My direct involvement in personnel matters was not just limited to senior officers but I tried to keep at arm's length from my very competent staff.

Also on the personnel side of the house was what was known as Corps Domestic issues. This covered a variety of subjects including, in no particular order, overseeing the approval and design process for the Princess Anne Banner, facilitating the work of the Corps Committee, ongoing review of the Corps Memorandum, progressing Volume 1 of the Corps History (which had been painfully slow), support for Corps Funds and some matters relating to dress and embellishments.

Under embellishments, we went through the process of introducing stable belts, identical to those worn by Royal Signals, to be worn with certain items of dress. These belts were not universally popular but they were eventually adopted throughout the Army although they were phased out a number of years later.

During my time as Director, there was great argument about whether or not officers should be permitted to wear berets – the main reason being convenience when travelling. Berets of various colours were traditionally worn by Armoured Corps (black), SAS (sand), Commandos (green), Airborne (red), Aviation (light blue) and Provost (bright red). Soldiers of other than those Corps mentioned above worn a navy blue beret.

At the height of the 'beret issue' I thought to play a joke on the Director of Artillery, Colonel Don Quinn. I had a spurious signal delivered by hand so it didn't go through the message system purporting to come from the Army Dress Committee approving use by Artillery officers of a white beret with red and blue pom-pom for ceremonial use and a blue beret with red pom-pom for day to day wear. The signal included the official codes for the various colours and had a real look of authenticity. As it happened, Don Quinn was out of town when the signal was delivered and his SO1 'buttons and bows' retransmitted it to the whole of the gunner world with a statement that it was false. Unfortunately for me that signal reached the Deputy Chief of the General Staff, Major General Peter Falkland, who had been a gunner and I found myself with heels together in his office being berated for misuse of the signals systems!

* * *

Important as this work was, I found that more and more of my time and energy was being spent on operational and equipment matters.

A new directorate had been formed in Operations Branch of Army Office entitled Directorate of Special Action Forces under Mike Jeffery. I had been exposed to much of the work of the UK SAS while in

London and had some peripheral involvement while in the JCB. With this background, I became involved in a number of SAS projects and worked closely with Mike, mainly on communications issues but also on some more general materiel requirements. I had multiple trips to the West to visit the SAS which I always found fascinating.

While in the JCB, I had some involvement with the Joint Facilities in North West Cape. I had little to do with this facility as DCOMMS but I did have some interest in the Joint Facility in Nurrungar. This facility was of major importance to space based surveillance and I was anxious to arrange for selected RA Sigs officers to become involved with the satellite aspects. After a successful submission to the Chiefs of Staff Committee, approval was granted and a programme of training in the USA with an attachment to follow was set in place. I was able to visit the facility and was greatly impressed with the work being done there.

There were multiple communications projects either in train or in advanced planning stages at the time including DISCON, RAVEN, PARAKEET and various EW proposals that were making great demands on the defence vote and there was a constant need to justify the expenditure. It really was unfortunate that so many of these projects were active at the same time as it did bring about a distortion in defence spending. These projects enabled me to renew many contacts in the communications industry, particularly with Racal, Plessey and Siemens which all added to my enjoyment in the job.

* * *

I was in Melbourne on 5 July 1980 to attend a parade where 126 Signal Squadron would exercise its Freedom of Entry to the City of Box Hill. I was to be received on parade by the Officer Commanding, Major Sam Smalley. All was going well until I attempted to exit my car. I wasn't all that used to exiting cars wearing a sword and I managed to catch the scabbard on the door sill. In attempting to correct my fumble, I knocked my cap off. It is very difficult to maintain one's dignity while trying to replace a cap with one hand while fighting with a scabbard with the other – not helped by all this occurring in the full sight of the troops on parade. I did recover but many of the soldiers had great difficulty keeping a straight face during my inspection.

The morning after the parade, I was in the Mess when I heard the news that Major General John Williamson, the General Officer Commanding Field Force Command, had died suddenly at his married quarter in Victoria Barracks, Sydney. This was shattering news for the Army and for the Corps as 'Big John' as he was known was seen as a prime contender for Chief of the General Staff.

The funeral for the General was held in St Stephens Uniting Church, Macquarie Street, Sydney followed by cremation at the Northern Suburbs Crematorium. I had the honour of representing the Corps at the funeral and was, because of the General's affiliation with Signals, afforded the status of a principal mourner.

The funeral was one of the largest seen in Sydney in many years as it involved a band and a battalion sized guard along with the many cars in the funeral procession. To give you an idea of the size of the funeral, Macquarie Street was closed for about thirty minutes after the service to allow the procession to clear the church. The cortege then had to take a circuitous route to the Crematorium to allow the guard and Artillery saluting party to get in position. I doubt if such a spectacle will be seen again in Sydney.

To mark the General's contributions to the Country, Army and Corps, an award, the JI Williamson Award, was established in 1982. The award is made to the outstanding student of the Regimental Officer Advanced Course and ranks second to the Lionel Matthews Award.

* * *

The death of John Williamson had both immediate and long term affects on the higher echelons of the Army. An immediate affect was that my old friend Major General Ron Grey was posted to Sydney to be GOC Field Force and Major General Peter Gration became Chief of Operations. In the longer term, one could speculate on what might have happened if Williamson had replaced Lieutenant General Dunstan in February 1982 and had stayed in the appointment for three or four years.

* * *

It became very obvious that the 'trial' separation would become permanent and Yvonne advised me she would file for divorce towards the end of 1980. I had had a number of discussions with my old friend Monsignor Gerry Cudmore who was the Army's Principal Catholic Chaplain about my situation and the unwelcome prospect of remaining a bachelor for the rest of my life. I was aware that there were circumstances where an Annulment of Marriage might be granted but I doubted if I would meet the requirements. Gerry was more hopeful than I and he put me in contact with Father (later Bishop) Pat Power who ran the Canberra Catholic Marriage Tribunal. I was very careful not to consult the various books on the subject on grounds for annulment as I wanted to present my case as honestly as possible. Father Pat read my submission but was not terribly hopeful. Nothing more could be done at that time as no action on annulment could be taken until a civil divorce had occurred.

Yvonne submitted her case to the Family Court in September and a divorce was granted in November 1980. I saw Father Pat again and then set about finding witnesses, etc for interview by the tribunal. I knew the process was a long one and I knew my chances were slim so all I could now do was wait.

* * *

As I mentioned earlier, I spent much time with Mike Jeffery working on various SAS projects. He was aware that I had failed a parachute course many years before and he offered to find a place for me on a water jump if I was so inclined.

I was so inclined and departed for RAAF Williamstown. We were a mixed bag in the RAAF Mess that night with all of us suffering various degrees of nerves. In particular, I remember a number of Fleet Air Arm crew who were decidedly unimpressed at being made to 'volunteer' for the jump. I must admit to sleeping very little during that night and wondered why I had accepted Mike's offer.

Morning came and I was delighted to find that the Officer Commanding 130 Signal Squadron, Major John Schmidtchen, had turned up to jump with me and provide moral support. He also offered me the hospitality of his Mess after the jump. Although he was a very experienced parachutist, he seemed almost as nervous as I was but I'm convinced this was a bit of an act.

Training for the jump was very basic and lasted for about an hour. We were using modern parachutes using Capewells. The Capewell lanyard-type parachute canopy release is a hand-activated mechanical device for detaching the parachute harness from the canopy. Its simple, easy operation offers paratroopers quick, safe disengagement from the canopy and was a vast improvement on the type of harness that was in use on my original course.

We were not using reserve parachutes, so instruction was limited to schooling us on the exit technique from the back ramp of the Caribou aircraft and the procedure to be followed to release the harness. From the best of my memory, we were told to uncover the Capewell at about 150 feet above water, thumbs in the harness release rings at 40 feet and pull away on hitting the water. As it is not possible to judge height above water we were told to take our reference marks from the top of the hills around the bay and the top of the trees growing close to the edge of the bay, respectively.

It was not long before we were loaded on to our Caribou for the short flight to the jump location at Shoal Bay. In what seemed to be a matter of a few minutes, we were circling at the jump height of 1500'. I cannot remember the order in which we jumped, but I was in the first few. At last the time came and I got to my feet, had my equipment checked by the jump master and shuffled towards the ramp. The exit technique was extremely simple-one step with the left, one with the right and the next left was out the door.

I was standing ready to go and got my tap on the shoulder. My feet were ready but my hand holding the cable wasn't. The jump master, not too gently, unwrapped my fingers from the cable and set me on my way.

I knew I had left the aircraft due to the absolute silence. I opened my eyes, looked up to see that my parachute had opened without any tangling. This was obviously very easy and I wondered what all the fuss was about. Trained parachutist (in my eyes) Evans was descending. Arms extended above my head with a firm grip on the rigging and enjoying the trip. 150' – hands down, remove Capewell cover and hands back up on the rigging. 40' hands down to find I had only uncovered one Capewell. Frantic activity, thumbs in the rings just as I hit the water and released the parachute. Unfortunately, I had not taken a deep breath and I wasn't prepared for the depth reached before starting to struggle upwards to the light. I thought I might drown but all I got was a lung full of water and emerged coughing and spluttering to be dragged from the water by the recovery team.

Notwithstanding the odd moments of terror, quickly supplanted by simple fear, the whole experience was very enjoyable and I certainly understand why so many people are attracted to parachuting as sport.

I returned to Canberra via Victoria Barracks where I was able to report my jump to General Grey and enjoy his hospitality before returning to my duties.

* * *

It was always busy in the Directorate and I spent a high percentage of my time visiting units and taking part in commemorative parades and social functions. I visited the School of Signals every two months or so to keep abreast of Signals policy development and to keep an eye on training standards.

Another function, and an important one, was to represent the Corps at the CGS Exercise and at various other national and international conferences. Like my predecessors, I felt it extremely important to maintain the profile and high standing of the Corps and, like them, looked for people who would be our best representatives in non-Corps and Defence postings and to put forward contenders for overseas staff and command colleges.

I'm not sure if I can take the credit for getting RASigs officers posted into Regimental Signals Officers positions in the Fighting Arms, but I certainly was a strong supporter of this policy. During the Vietnam War, the Corps had not always (and I emphasise always) put our best officers in such appointments and we suffered for it for many years after the war ended. This policy was not always popular with our young officers and some of them needed convincing that we looked for our best to be RSOs. As 1980 drew to a close, I considered that the Corps was in good shape and I was getting some feedback indicating I was doing a good job.

* * *

I was no sooner back at work in January 1981 when I was interviewed as a candidate to attend the Advanced Management Programme at the Administrative Staff College, Mount Eliza to be conducted from 25 March to the 8th May. I had no real knowledge of the College although I did understand the course was highly regarded in both military and civilian circles. I'm not sure how many candidates there were but, for better or worse, I was selected to attend.

The two months between selection and attendance were as hectic as usual and I was looking forward to the break and the possibility of indulging in some reflection. During this period I moved out of Brassey House into a flat and was just able to get it into some order before heading South in my restored MG.

The College itself is housed in 'Moondah' a grand old mansion set in beautiful grounds with student accommodation discreetly out of view. The accommodation was reasonable enough although our numbers dictated two to a room – not the most satisfactory of arrangements. Dining facilities were good and we had access to a bar.

There were sixty-one students in the course distributed among six syndicates which did not change through the course. The student body was a mixed lot with the majority being from state-owned or commercial industry. There were seven overseas students, a clergyman from the Wesley Mission, one police officer and two servicemen – Wing Commander John Townsend, RAAF, and myself.

Most attendees held senior positions, particularly those from industry.

What I found interesting was the pains that the Principal, Emeritus Professor Bill Walker, and some, but not all, of the staff took to promote the reputation and standing of the College and the idea that becoming one of the *collegium* would open doors of opportunity all over the world. I was not alone in thinking there was considerable hyperbole involved in the protestations but we were all of a mind to embrace the programme with enthusiasm.

As one would expect, one of our earliest projects was to give an autobiographical presentation to the staff and student body. This was taken by most to be a simple exercise but we had one student, a senior industrial executive, who shall remain nameless, who became very agitated at the prospect. He told some of us that he had never ever been required to give a presentation to his peers. He had no trouble briefing his seniors or directing his staff but peer review threw him completely. He was so agitated, he obtained some tranquillisers from his GP and then fortified himself with the odd Johnny Walker before mounting the podium. His presentation verged on the incomprehensible but we all applauded with vigour which restored his equilibrium.

In the discussions that followed these presentations, it quickly became evident that civilian enterprise spent very little on the education and training of management staff. In fact, the majority of attendees acknowledged that this course was the first in their careers that had been paid for by their employers. In contrast, at this point in my career, I had followed four years at RMC with two years full-time doing my BE, two years part-time doing my MEngSc, one year on satellite communications training, one year at Staff College and six moths at the

Joint Services Staff College. It was very easy to conclude why military officers were so attractive to industry.

The Directing Staff were also a mixed lot. Most had impressive qualifications but some seemed to have had little experience in actual management positions. This seemed to me to be a shortcoming of the College. Notwithstanding this comment, I found the close contact with the student body to be extremely valuable.

I also found attendance at some optional sessions given by one of the resident psychologists very helpful at a personal level. I was enjoying my career but was finding it increasingly stressful. At the same time, I was not very happy. My marriage breakdown had been rather traumatic and I was not coping well living the bachelor's life. At this stage, I was not very hopeful of obtaining an annulment and I did not think I could bring myself to marry outside the Church. I suppose I was suffering a form of depression but had not discussed this with anyone.

I can't remember the actual title of the optional sessions, but they were directed at contending with stress and related topics. I was surprised by the numbers that attended (at least a third of the students) but I guess this was to be expected given that most of us were in the mid- to late-forties. Anyway, I found the sessions most helpful and was in a better frame of mind when I left to return to Canberra.

All in all, I found attendance at 'Moondah' valuable and I did keep in touch with the Alumni association for some years but this lapsed after a few years only to be rekindled many years later when I joined industry after retirement.

* * *

During the rest of May, June and July I had a particularly heavy travel programme which included the 21st Birthday Parade for 1 Signal Regiment, Cabarlah, Adelaide, Townsville and Hobart along with my regular visits to the School of Signals.

It was also during this period that my involvement with St John Ambulance picked up and I became a member of the St John Council in the ACT. I was still a regular attendee at the Jazz Club in Dickson.

I was at the Jazz with Shirley at the Dickson Hotel on the Saturday of the 1981 Queen's Birthday weekend. As we left the hotel there was a notice saying that the Queen's Birthday Party would commence there at 19:00. Shirley and I decided to attend so we went over the road for a meal in the Malaysian restaurant before returning at the appointed time. We received a mixed greeting – Shirley was not allowed to enter but I was told in a simpering voice that I would be most welcome. Having now realised that it was indeed to be a "queens" birthday party, we retreated to Garran.

In early August the Chief of Materiel made me an offer I could not refuse. He asked if I would be willing to lead a delegation on a world trip to look at technology for Project Parakeet, the new area communications system. I was to be accompanied by Lieutenant Colonel Ross Thomas from Materiel Branch, Rex Shoubridge from Defence Science and John Davies who was working in the department overseeing industry development in major projects.

Before going away, I made a visit to the School of Signals and while there met Major General Kevin Cooke who was the GOC 3 Division to discuss communications issues in his division. I was very committed to working with the Army Reserves as I was well aware of the dedicated service given by part-time soldiers.

Back in Canberra, I had the pleasure of being present at the wedding of Shirley's eldest daughter, Linley, to Dr Deryck Scarr, an academic from ANU.

* * *

We left Sydney on the 22 August with an ambitious programme that would take eight weeks and have us visit all the main players in area communications. The programme included visits to Army units; government laboratories and testing grounds; civilian laboratories and factories in addition to the various military staff organizations involved in the of procurement policy and tactical employment of area systems.

We had an inauspicious start to our journey. We were booked on a flight that would take us from Sydney to Los Angeles via Honolulu allowing some rest time on arrival before our first engagement. We got caught up in a QANTAS problem that required our aircraft to divert to Fiji to deliver an engine. The delay caused a missed connection and we ended up in San Francisco with about half an hour to collect baggage and join a flight to Los Angeles. By the time we were settled in our seats on this last leg, we had been in the same clothes for over 24 hours and were both tired and sweaty. I ended up in the near empty huge front compartment of an American Airlines Tri-star. As soon as we were airborne, I called the stewardess, a very tall, statuesque African American, and asked if I could smoke my pipe. She indignantly responded by telling me that not only was I in a non-smoking section but drug taking was strictly prohibited. She had thought I had asked if I could smoke some pot and was not mollified until I produced my pipe and tobacco pouch and repeated the request. The stewardess was greatly embarrassed by the misunderstanding and with much laughter waived the smoking prohibition and produced a large paper bag with enough miniature bottles of alcohol to keep us in drinks for most of our time in the USA.

Our first visit was to Singer Librascope, one of the major contractors for the US TRITAC System. The same afternoon we boarded a flight to Tucson, Arizona to visit the US Army Proving Ground at Fort Huachuca. From there were turned to Tucson and thence to Ford Aerospace, Colorado Springs. Then to Dallas to visit Collins Radio before flying to Ottawa to Canadian Defence Headquarters. A short trip then to Quebec to the Defence Research Establishment, Valcartier. Back across the border to Raytheon at Boston then ITT at Nutley, New Jersey and then a little further South to visit the TRITAC Office at Fort Monmouth. After Fort Monmouth we went North to New York City for a few days off and collate material collected to date.

While in New York I decided that Shirley was the girl for me and on the 7th September I telephoned to propose. She responded tenderly with the question *"Are you sober?"* I said I was but she insisted I call again in the morning – I did so we were engaged. I asked what sort of ring she would like and was told Ruby – more on this later. From New York we went to Camden, New Jersey, to visit RCA and then drove on to Washington to call at the Pentagon and then the Embassy to make courtesy calls and despatch documentation back to Australia. Our final port of call in the USA was to Fort Gordon in Atlanta, Georgia which had become the home of the US Signal Corps.

We had spent three hectic weeks in the USA and Canada and I think all of us were feeling a little drawn and weary. At the outset of the trip we had discussed the fact that we would be constantly in each others company and had decided that as much as possible we would seek to be separated on all aircraft journeys. This had worked well and there were no real tensions in the group, just the odd irritation quickly resolved.

Wherever we went in North America we were well received, provided with all the information requested and shown great hospitality. We had anticipated this and took with us a good supply of presentation plaques, ties, etc and just as well as we had to offload many mementos at the Embassy on our way through.

I should add that for most of our trip we were accompanied by Lieutenant Colonel Bob Denning from the Embassy staff who had made extensive contacts and did much to facilitate our trip.

The next leg of our journey was to London where we arrived on the 12th September for a rest day. A valuable day was spent at the Ministry of Defence followed by a day at the Royal Signals Research Establishment at Christchurch.

When it was settled that I would be on the trip, I arranged for the Colonel-in-Chief of the Signal Corps of the old Commonwealth, Princess Anne, to be advised that I would be in the UK for the week commencing the 14th. This generated a flurry of messages in which the Princess expressed to wish to receive me. Initially, I was to call at her country residence but this was later changed to London. Waiting for me at Australia House was a letter from the Private Secretary to HRH confirming that I would be received at Buckingham Palace at 11:00 on the 16th.

So instead of visiting the School of Signals at Blandford (Ross Thomas headed up the delegation in my place), I prepared myself for the audience. The High Commission was extremely helpful and provided me with one of those enormous Daimler limousines for me to be delivered to the Privy Purse Door where I was met by the Private Secretary, Major Lawson and a Page wearing breeches and lace ruffles, etc. I tried hard to remain calm while walking along what seemed like a mile or so of corridor lined with paintings and fine furniture. We eventually arrived at HRH's apartment and I was ushered in and presented by Major Lawson. I was put at ease immediately and then spent the best part of an hour in pleasant conversation. Princess Anne showed a real interest in the Corps as it existed and what was planned for the future. We also talked about her charity work and interest in equestrianism. The whole affair was a quite delightful experience that I have long cherished.

Then back to work with visits to Cossor, Racal, Decca and Plessey before leaving London late on the 21st for a three day visit to Paris for discussions with government and Thompson CSF and demonstrations. We were quite amazed on the Wednesday when the entire 18 Signal Regiment were put into the field at Epinal to demonstrate PRO-RITA. This was a good example of the close cooperation between government, the military and industry when there was a prospect of a large contract.

From Paris we went to Siemens in Munich and from there to Amsterdam to look at the Phillips SIGNAAL system. From there we went to Oslo to visit the Norwegian Military HQ and then a day with Elektrisk Bureau. Our last stop in Europe was at Stockholm to look at the Swedish Single Radio Access system.

We returned to London and had a free day before heading home via Singapore. I used the time to make a call on the International HQ of St John Ambulance at St John's Gate to pay my respects and bring greetings from the ACT.

We broke our return trip at Singapore with a rest day. And now I return to the ring. I made the mistake of taking Ross Thomas with me when I visited a jewellery store that had been recommended. I asked to see their selection of ruby rings with Ross looking over my shoulder. I made my first selection to hear Ross' comment "*cheap Charlie*". This occurred for my next two attempts until finally I met with his approval. I will not divulge the cost but it was very much more than I had intended!

The brunt of the work preparing the report of the trip was laid on Ross Thomas. As expected he did a great job and the report was well received by the Chief of Materiel.

* * *

The last months of 1981 were more or less a repeat of 1980 with a round of Corps Committee meetings, visits to units, parades and the November round of Corps Dinners.

I had heard from the Marriage Tribunal and had a number of meetings with Gerry Cudmore. I was advised that the Regional Tribunal had approved the annulment but this had not been accepted by the National Tribunal and I was required to appear for another interview. I was greatly encouraged at the second interview but I still ended the year with some uncertainty.

I was approaching the end of my appointment as DCOMMS-A without any real idea of what might be next in store. I had been called to the Military Secretary's office for the first of a number of meetings to discuss my replacement. Eventually it was decided that it would be Phillip Skelton on promotion with a hand-over in February 1982.

Commodore Harry Adams had been appointed as DGJC-E and he was likely to remain there until the end of 1984. I had thought that if I was to be promoted it would be into the DGJC-E job so I would probably have at least one more appointment as a Colonel. One prospect was that I might replace Colonel (later Brigadier) John Moyle as Director of Protocol and Visits. I wasn't too keen on this idea but it was the only appointment on offer at the time.

Very early in the New Year I heard that the Engineering Development Establishment (EDE) was to revert to being an Army unit after ten years as one of the Defence Science laboratories. Out of the blue I had a call from Major Colin Boughton who had been posted to the Military Secretary's Office who told me a panel was being prepared to select a new Head at one star level. I told him I would be very interested and left it at that.

No interviews were held and I heard nothing more until I was called by the Chief of Materiel, Major General David Engel, to say I was to be promoted to brigadier on the 26th of February and I was to take over as Head of EDE on the 10th of March.

I found my two years as Corps Director extremely rewarding and enjoyable. I had the great advantage of a very professional staff who continually performed their duties with diligence and enthusiasm – what more could I have asked? I've heard it said that all a CGS (or Chief of Army) can do in a two year appointment is change one item of dress. This is no reflection on the officers concerned but is an indication of the power of inertia in a large organisation.

Well, what did I achieve? Lets get the item of dress out of the way – the stable belt, but this was reasonably short lived anyway. Corps profile and standing? – I thought I maintained these. I certainly can claim the credit for getting approval for the Princess Anne Banner. I also think that I was able to identify high quality officers, senior NCOs and Warrant Officers and facilitate their further development and progression. However I was regarded, I certainly could not have put more time and effort into the job than I did. I thought I had done a reasonably good job as Director but I will leave it to others to cast their judgement.

ENGINEERING DEVELOPMENT ESTABLISHMENT

EDE derived largely from the establishment of design elements within the Master General of the Ordnance Branch on 13 October 1939 and this date is the basis for anniversary celebrations within the unit. I do not propose to go over the history of EDE as this is well covered in a publication 'Army Engineering Capability – A History 1939-2000' compiled by Margaret Kaczmarek. I will mention only that the control of the Army Design Establishment passed to the Defence Science and Technology Organisation on 1 December 1974 and was renamed the Engineering Development Establishment in recognition of the unit having tri-Service focus. John Wisdom was appointed Head EDE and he reported to the Controller, Service Laboratories and Trials, initially Air Vice Marshal R. Noble.

Transfer of EDE to DSTO was not universally successful. There was never any doubt about the technical competence of the staff or the quality of its output. There was, however, growing disquiet about the responsiveness of the establishment. EDE had become know as the Engineering DELAY Establishment by all and sundry with resultant damage to its reputation. Responsiveness became a key issue which drove the move to bring EDE back under Army control. Another consideration was that EDE was small 'r' and large 'D' and it did not fit easily into the scientific communities of Defence Science. In later conversations with members of staff, I found that most of the professional staff welcomed the move back to Army.

John Wisdom had retired at the end of 1980 and for 1981 Harry Brekell and Rex Christensen took it in turns to act as Head. Both were long-serving members of EDE and both quite reasonably saw themselves in contention for the appointment. At the time of my appointment, Rex Christensen, who I had known since MALLARD days, was acting Head but I had never met Harry Brekell.

In mid-February 1982, Phillip Bennett was promoted to Lieutenant General and appointed CGS. I had a meeting with him before moving South during which he assured me of his support and promised to visit EDE at a suitable time later in the year.

I completed my handover to Phillip Skelton and he assumed as DCOMMS-A on 26 February. I moved into an office in Materiel Branch and commenced a series of briefings on current projects and the task review process. During this time I also met with Professor Tom

Fink, Chief Defence Scientist, to discuss future cooperation between EDE and the DSTO laboratories.

Before leaving Canberra, I had a final briefing with Major General Engel to receive my 'riding instructions'. These were direct and simple and were music to my ears.

"You have your Task List and you have your budget so get on with it. I will not be looking over your shoulder but come to me if you need support."

His final direction to me was that I should come to Canberra once per month for lunch and an informal chat about how things were going.

With these instructions, I packed my bags and flew to Melbourne.

* * *

My first problem was to find somewhere to live. I had the option of living-in at the Watsonia Area Mess but I thought this would not be fair to the incoming Corps Director. I didn't fancy setting up in a flat until I was settled into my new job and, as there was no suitable military accommodation in the Melbourne area, I moved into the Naval and Military Club.

My first day was the 11 March. My staff car arrived at the appointed time and we headed into the wilds of Maribyrong. The Acting Head, Rex Christensen, was away but I was met by Harry Brekell, escorted to my office and introduced to all the senior staff.

My office deserves a mention. Entry was made through a small room occupied by my secretary. Another door opened into the office proper. The first item was a coat cupboard and book shelf. Then came a conference table with eight chairs followed by a coffee table and four arm chairs and finally my desk. Along the street side were two large windows with no view and a distinct disadvantage that I will mention later. It really was most impressive and about four times larger than any that I had occupied previously.

When I took over, the total strength of EDE was close to 400 civilians and 20 soldiers. The organisation was quite simple with Mechanical (Harry Brekell), Electrical (Rex Christensen), Engineering

Facilities (Noel Olver), and Administration Divisions (Kevin O'Sullivan) together with a Planning and Coordination Division headed by Lieutenant Colonel Mike Croft. The EDE Workshop facility was located in an old remount depot building some hundreds of metres away across a busy intersection and the Trials and Proving Wing was located at Monegeetta, located between Sunbury and Romsey. There was an Officers Mess quite nearby which was open to senior members of staff.

The appointment of a one-star to EDE had not been entirely expected and my Ford Fairlane was not yet available. As an interim measure, the system provided a red sedan (I think it was a Holden) complete with a One-Star plate. The car was first seen in public when I made a call on the Commander of the Third Military District. Unbeknown to me, my driver had responded to a question from another driver about the red car by stating there was a new policy in place to put Red Hats into red cars. I had many irate calls on the subject but it added some comic relief to my early days in the job.

When I took over as Head EDE, I knew there would be many challenges and I was determined to avoid any precipitate action that might have unintended consequences. Looking inwards, I had to establish my credibility with the staff and then proceed to do what was necessary to improve responsiveness. Looking outwards, I saw my main job as restoring the reputation of the establishment and promoting it to the Army as being well worth the money spent on it.

My academic qualifications were as good as any of the professional staff but my technical expertise was mainly limited to communications-electronics. So I set about spending as much time as possible in acquainting myself with all the current projects by talking to the project leaders and support staff. In particular, I had to come up to speed on small arms and mobility technologies. I needed to not only learn the nuances of each project but also to convince the staff of my genuine interest in them and their projects. I needed to be sufficiently knowledgeable to champion the case of EDE against all outside agencies.

It took me a couple of months, but I found that, by that time,

I had acquired sufficient knowledge of the projects and the facilities to be able to take the lead in conducting important visitors through the establishment. When doing so, I was also at pains to ensure the responsible project officers had an opportunity to explain the status of their projects.

All members of the staff were aware of the reputation EDE had, rightly or wrongly, acquired as a source of delay and I sensed that morale was not as high as I would have liked. So this became another priority for me – to address the whole staff at every conceivable opportunity to try to convince them of my high regard for the establishment and of my driving ambition to lift the profile of EDE as a valuable asset for the Army in particular and Defence in general. To assist me in this I took every opportunity to invite senior members of the Army to inspect the facilities and be briefed on projects of interest and importance. I was receiving a great deal of encouragement in my new appointment and was delighted to welcome the CGS and CMAT to EDE in April 1982.

I was well aware of the reputation for delays in projects so I set about trying to determine just where the choke points were. Clearly one of these was the drawing office where there was a considerable backlog of work. The reason for the backlog was not immediately obvious so this became an area of interest for me and my staff officer. After a month or so, it also became obvious that there seemed to be a major problem in bringing projects to a conclusion. Project officers always seemed keen to extract the last ounce of value from their work and this invariably introduced delays without necessarily making a significant improvement in project performance.

As you will gather, I was kept very busy but I was also finding the appointment immensely satisfying.

* * *

In mid-March I received the very welcome news that my annulment had been granted and that I was now free to remarry in the Catholic Church. So Shirley and I were able to start planning our wedding. 14th

August was selected as the date and arrangements were made for the wedding service to be held in the RMC Chapel at Duntroon with the joint celebrants being Monsignor John Hoare and Shirley's priest, the Reverend Fred Hart.

This also meant that I needed to find us a place to live and I started the search for a town house that would be conveniently placed for ready access to EDE and for Shirley to work in the city. I finally selected a town house in West Brunswick and took up residence in anticipation of Shirley joining me in August.

* * *

At the end of April 1982, I attended my first Task Review Conference. This involved a complete review of all current projects and was attended by all the relevant Materiel Branch project managers. During the conference each of the EDE project managers would report in detail on progress, percentage of task completion, significant problems, projected completion date, etc. The atmosphere was confrontational and, in my opinion, was not conducive to cooperative problem solving.

In my preparation for the conference, I examined in some detail statements by project officers on task completion. It seemed the usual practice was to report on the amount of effort expended as a percentage of the estimated effort required to complete a task. This had little to do with enabling an accurate assessment of when a task would be completed and so compounded the idea of inordinate delays at EDE. The basis of the problem was that it was virtually impossible to estimate, at the start of a project, the amount of effort that would be required to complete it. I had no idea how to overcome this problem but I was determined to improve the tenor of future conferences.

Not long after the conference, I had my first meeting with the heads of divisions to discuss the next years budget. It soon became obvious that in the past budget allocations were made without reference to need. Overheads, including the running of the Facilities Division, were taken out and the remainder simply split equally between the two engineering divisions. Some little bargaining took place and then someone suggested the budget be put to the vote. I made it clear that the budget was not subject to voting but would be my decision based on divisional submissions.

It was at about this time that I made the decision to spend some money on the small theatre that was used for visitor briefings to improve the audio/visual facilities, lectern facilities and the like. I also commissioned a professionally made audiovisual package featuring EDE's staff, facilities and current projects which could be used as a general introduction for visitors before specific briefings. The package was configured to allow it to be used outside the establishment. There were some in EDE that thought this was a bit of an indulgence but it soon became obvious that it was a positive move to enhance our standing.

* * *

In August I took a month off and returned to Canberra for the wedding. The wedding itself was reasonably low key – no uniforms nor Guard-of-Honour but with a fair number of guests. Jim Messini was my Best Man and Shirley's Matron-of-Honour was Rae Mann, one of her oldest friends. We had our reception at the Hellenic Club with my classmate, Steve Hart, as MC.

For our honeymoon we flew to Noumea for a week and then had a week at the Erakor Resort on Vanuatu. While at Erakor we met the local village head man who controlled the resort workforce. He invited us to join him at their Sunday church service and have lunch afterwards. The church service involved much singing and was very joyful but, to us, unusual as the men and women sat on opposite sides of the church.

After church we went to the chief's house where, as a special privilege, Shirley was invited to eat with the men. I was the guest-of-honour so was served with the native delicacy of a large fish head cooked in coconut milk! I must say that I had to wield considerable will power to work my way through the meal.

The return trip was to be via Nouméa. When we got to the terminal, we were advised that there had been an error which had led to overbooking and some passengers would have to remain another night on the island. This did not present a problem but I approached the booking staff and asked if a message could be sent to my mother and stepfather as they were collecting us at Mascot after our honeymoon. 'Honeymoon' did the trick and Shirley and I were upgraded to First Class.

As we waited on the aircraft to take off a steward appeared offering champagne and orange juice. Shirley accepted but I asked for mine straight. This was misinterpreted and I was given a glass of orange juice. This was regarded as a great joke with the story being repeated regularly at family gatherings.

We returned to Melbourne to take up residence in West Brunswick and Shirley commenced work with Veterans Affairs in St Kilda Road.

* * *

Throughout the remainder of the year I spent time traveling to meet with our 'customers' and to liaise with the Defence Laboratories and the various firms that were under contract to provide field equipment for the Army. I also managed to keep in close contact with the Corps of Signals by regular visits to the Mess at Watsonia.

Throughout 1983 there was an increasing number of visitors to EDE which I always welcomed. After some encouragement, EDE was added to the visit list of Staff College, JSSC, the Industrial Mobilisation Course and various Army courses. We also had the pleasure of entertaining many overseas visitors from various military laboratories. In addition, there was a constant stream of contractors' representatives discussing project-related matters. I was absolutely convinced that the money spent on our briefing theatre had paid for itself many times over. In line with EDE being an Army resource, I offered the accommodation facilities of the Trials and Proving Wing at Monegeetta to Major General Cook at 3 Division. This was gratefully accepted as it allowed valuable CMF time to be spent on site rather than on traveling. Similarly, the climatically controlled test house was made available to the Army Medical Corps for some testing on health issues related to climate.

Towards the end of the year I attended a course at the University of New South Wales on the Management of Research and Development. I found this extremely useful, not only from the course content but also for the opportunity to compare notes with other managers of laboratories. I had already been looking at ways to enhance the effectiveness of EDE and the course added impetus to a study I was conducting on its organisational effectiveness which I concluded in 1984 – more on this later.

I considered 1983 to have been a successful year. Morale at EDE had been markedly improved and there seemed to be an improved level of cooperation between ourselves and our Materiel Branch masters. A number of projects were well on the way to completion and the use of the pejorative term Engineering Delay Establishment seemed to be on the decline. From my point of view, 1983 had been a year of consolidation and I was ready to examine new organisational arrangements and ways of operating in 1984. But before doing so, I sought and gained approval to visit a number of establishments similar to EDE in North America and the UK.

* * *

I had devised a visit plan that would allow me to not only examine organisational and management issues in like organisations but also to allow me look at developments in armoured vehicles and to have discussions in the UK with Plessey and Racal.

After courtesy calls with the Head of Defence Staff and Australian Army representative at our Washington Embassy, I spent the best part of a day with the US Development and Readiness Command. Then to Fort Monmouth to visit the TRITAC Office, Communications and Electronics Command and the Satellite Communications Agency. The following Monday, I spent a day with ARADCOM, the Army Air

Defense Command, to look at air defence missile development before heading to Boston, where I met up with my old friend Jim McKeon.

In Boston, I visited the Natick Laboratories also known as the US Army Soldier Systems Centre. The Natick Laboratory was very similar in size to EDE and so was of considerable interest to me from the management and organisational aspects. Natick responsibilities included, inter-alia, research on textile for combat uniforms, body armour, combat rations and parachute development including parachute delivery systems.

I had one embarrassing moment during my tour of the facilities. I noticed a chromed device about the size of a large cabin trunk fitted with many pipes and meters. I was intrigued and asked just what was its function. The response *"It's our newly developed combat ice-cream maker"* caught me by surprise and I was not able to stifle a short laugh. This caused some indignation and I was told in no uncertain terms that US soldiers were entitled to their ice-cream whatever the circumstances. I was duly apologetic and my tour continued without more ado.

My next visit was to Tank Automotive Command located in Warren, Michigan. For this visit I was accompanied by Major Peter James from our Embassy. The Abrams main battle tank had been developed here and in-service development was still a major task for TACOM. I received a number of briefings on the Abrams and had the opportunity to have a short ride. I was also briefed on a new project – the Mobile Protected Gun Platform (MPGP). As the briefing went on I stopped the briefing officer to ask a series of questions:

"The MPGP is armoured, right?"

"Yes, Sir."

"It's tracked, right?"

"Yes, Sir."

"It has a gun mounted in a rotatable turret, right?"

"Yes, Sir."

"Then why don't you call it a Tank?"

"We can't, Sir. The Abrams is still in development and Congress would not approve funding for another tank."

This reminded me of the way that the US Army had to stop spending money on High Frequency Radio in order to get funds for development of satellite communications.

From Detroit, I next went to Ottawa to visit the Defence Research Establishment Ottawa; the Canadian Quality Assurance Organisation and the Land Engineering Test Establishment (LETE). I was particularly interested in LETE due to its similarity in size and role to EDE and I suggested that it might be worthwhile arranging an exchange programme between the establishments. Unfortunately, this suggestion never got off the ground.

The next leg was to London for a hectic week of visits to the Ministry of Defence, Racal, Plessey, the Military Vehicles Engineering Establishment, the Royal Armament Research and Development Establishment and the Royal Signals and Radar Establishment.

One aspect of this visit that was quite exciting was the interest being shown by the Electronic Warfare staff of the MOD on the Australian airborne single station locator.

Overall the trip was most successful even though it was very rushed.

* * *

I now felt it was time to do a thorough review of EDE's establishment. I had a number of concerns, the main one being what I will call 'reverse retention' – engineers rarely left EDE and there were many on staff who had spent their whole working life in the establishment and there was limited scope for progression. The other problem was common to all R&D establishments and that is that promotion tends to follow administrative ability rather than engineering excellence. Unfortunately there was little scope within the system to reward engineering excellence by pay rises or other 'perks' which meant advancement was based on becoming section leaders, then group leaders and finally divisional heads.

The other pressing problem was that the rate of change of technology was outstripping broad expertise. By this I mean that the two main divisions of mechanical and electrical engineering were no longer appropriate. I thought that one simple solution was to break mechanical engineering into armaments and mobility and electrical into communications and power. This had the added advantage of adding two new Engineer Class 5 appointments to the establishment.

Another issue was the growing tendency of EDE to be more involved with project oversight rather than actual design and development work. I had overheard many times 'outsiders' saying 'what would you know about design – you never do any'. There was no organisational fix for this problem but I did introduce a policy that would allow engineering staff to spend some of their time in following their own lines of interest. I made it clear that I did not care what these projects were as long as they involved developing a design and producing a prototype.

I was somewhat surprised with the speed with which my proposals were accepted and a new organisation approved. Apart from the new divisions, new groups were formed covering Small Arms; Communications Architecture (PARAKEET); Armoured Fighting Vehicles (WALER); Command, Control and Computing; Software Systems; and Mobility Terrain Analysis. The changes did not occur overnight but were well in train by the end of the year.

* * *

Shirley and I were due for a holiday so we decided to go to Europe with the main purpose of attending the Passion Play at Oberammergau. Our intention for the trip was to hire a car, keep off the Autobahn, take easy stages and each afternoon look for a Zimmerfrei (Guest House). I had spent some considerable time looking at maps and tour guides and had a fairly rough idea of the route we would follow either side of Oberammergau. Shirley already had some German and I had spent some time in evening classes trying to learn the rudiments. So we felt reasonably confident that we could travel ad hoc.

We started our trip on 30 June flying with Lufthansa via Hong Kong to Frankfurt. After a day of rest and sight-seeing in Frankfurt we collected our car, an Opel Kadet which Shirley insisted on calling a motorised skate board, and headed South to Heildelberg.

As we entered the city I saw a sign to the US base, Camp Patrick Henry. I had a flash of inspiration and thought we would see if there was a Post Guest House on the base. They certainly had and we were shown to a two room apartment by an apologetic manager who was concerned that the rooms were not quite appropriate for my rank. We were more that happy with the rooms and the tariff, $20 per night.

The next day was the 4th of July so we entered into the spirit of this most important US holiday. While in the Officers Club, we were asked what our next destination was. When I replied Baden-Baden and the Black Forrest, it was suggested we call at the Canadian Base at Lahr. Lahr was a Canadian Air Force base protected by a battalion of Princess Patricia's Canadian Light Infantry which is affiliated with the Royal Australian Regiment.

When we arrived at Lahr, I presented my ID at the Guard House and enquired if there was a Post Guest House. We were asked to wait and within a few minutes a jeep with flashing blue lights appeared and we were escorted to the Station Commander by a young Captain who complained that we had not given much notice of our visit. I explained we were on leave and there was nothing official about our visit. We received a profuse welcome and were escorted to the Edinburgh Suite which was in a small building adjacent to the Mess. The suite consisted of two bedrooms, a kitchenette, a sitting room with a bell to press if we wanted anything from the Mess. The suite had been recently renovated and renamed for a visit by the Duke of Edinburgh. Along with the package of welcoming documents was a CX (Canadian equivalent of the US PX) Card which allowed us to buy fuel at some ridiculous price at the Post garage.

We used the suite as a base while doing some local sightseeing including a trip to Stuttgart to see the Mercedes Museum but, unfortunately, this was closed when we arrived.

We were royally entertained by the Canadians and when I went to pay my bill I was told that we were the guests of the Fighter Wing.

While making our farewells, I was asked about our future plans. I said that after Oberammergau, we would head to Salzburg. We were advised to try the US Army accommodation in Berchtesgaden and, in fact, a booking was made for us. At this stage I had spent only $40 on accommodation (Frankfurt had been prepaid) so now it was hotels all the way.

We went to Oberammergau via Ulm and Memmingen. Oberammergau was a beautiful town and the Passion Play was everything that I had expected. From here we went to Innsbruck and then to Berchtesgaden. We called at the US Army hotel but the manager said that the accommodation was unsuitable for a one-star and we were shown to the Evergreen Lodge, which had been the home of Albert Speer. The manager was apologetic and told us that, as Australia was not a part of NATO, we would have to pay a $25 per day surcharge. The lodge was magnificent with wonderful views of the Alps and surrounding countryside. While we were there we had a look at the Eagles Lair, built for Hitler but which he only visited a few times.

We next visited the town of Steyr and had a tour of the rifle factory. From thence we went to Vienna via the magnificent Benedictine Abbey of Melk. Our next port of call was Munich, my favourite European city.

One of the ongoing projects at EDE was a field cooker and a contender was Haas and Sohn. Before leaving Australia, I received a call from the H&S agent, Harry Weingart, who offered us accommodation in Munich and arranged a factory visit. I accepted the offer and we had an interesting stay. Harry said we should drive to the H&S factory some 520 km away. What a trip? We travelled in Harry's Mercedes 500 SC Coupe along the autobahn. It was a little frightening, especially when Harry started playing with his car phone while we doing over 200 kmph. He gave me a drive on the way back and I hit 210 kmph. While cruising at about 200 kmph, a Police Porsche came up behind us flashing its lights to let him pass.

We were now almost at the end of our trip so we had a leisurely drive to Cologne where we stayed in the very elegant Excelsior Hotel just opposite the Cathedral.

After settling-in, I went to return the hire car. The office was located in an out of the way location and when I finally got there, the office was closed for lunch. I started to stroll about and then noticed that I was surrounded by leather and bondage shops, a Gay Kino, a few dubious bars so I rapidly retreated to the car where I hid until the Car Hire office opened.

The following day we checked our baggage with the Lufthansa Express train which delivered us directly to the Frankfurt airport for the long trip home.

* * *

The remaining months of 1984 were as busy as usual and I still found the work interesting and challenging. By the end of the year most of the establishment and management arrangements had been set in place and I was happy with the changes I had made. My replacement was to be Brigadier George Salmon who was a gunner with a BE in electronics, a MSc in guided weapons and a wealth of experience in materiel matters. I thought he would be excellent in the job and was more than happy to be handing over to him.

I had found my time in EDE extremely interesting and immensely satisfying. I had had a great team of engineers and supporting staff and I believed that I had established good rapport with most if not all of them. I believed that I had improved the standing of EDE with the Army and I took some pride in the job I had done. I said on more than one occasion that I would return to EDE at any time and this did occur some years later.

For now, I had been posted back to Canberra to be the Director General Communications-Electronics and I was looking forward to the challenges that appointment would bring.

JOINT COMMUNICATIONS BRANCH

I assumed the appointment of DGJCE on the 21st January 1985 just two weeks short of thirty years after I entered Duntroon. I then held the senior communications appointment in the Australian Defence Force with an extensive range of responsibilities.

I had been extremely lucky. When I was DCOMMS-A, I had aspired to this appointment and thought, correctly, that it would become vacant in 1985 after Commodore Harry Adams completed his three years. I thought, if I was lucky, I might get the appointment on promotion. As it happened, through circumstances completely out of my control, EDE had returned to Army and I was selected to be Head of Establishment and, as a result, was promoted to Brigadier three years earlier than I had expected.

So here I am in Russell as I said at the outset, in my small and rather shabby office in Defence Headquarters with a commanding view of the rear car-park and the Russell Offices' boiler room. Thirty years have passed remarkably quickly but I now have more challenges to face.

This book has been in gestation for almost twenty years. It remained stagnant for a number of periods over that time and has only come to a conclusion through the encouragement of my old friend Major General Ian Gordon. I hope the reader might find it of some interest.

There may be a sequel *Another Half Mile In Twenty Years – From Russell To ADFA*, but for now I'm heading to my workshop to make some sawdust.

PART II

ANOTHER HALF MILE IN TWENTY YEARS

From Russell to ADFA

PROLOGUE

Several years have passed since *Half a Mile in Thirty Years* was published and I have been encouraged to continue the saga up to the time when I finally left the workforce.

Half a Mile in Thirty Years ended when I returned to Canberra to take up the appointment of Director General Joint Communications-Electronics. Before continuing with my adventures in that posting, I thought it worthwhile to retrace my steps and cover several side roads, which became important in my later years. These include my association with the Duntroon Society and the Venerable Order of the Hospital of St John of Jerusalem.

On 27 June 1980, the then Commandant of the Royal Military College, Major General Alan Morrison, re-established the Duntroon Society. There had been a society formed in 1920, but this did not survive as the 'establishment' saw connotations of a 'union'.

In 1980, the Society's first newsletter noted that General Morrison had long held the view that there was a need for an organisation through which former cadets and past and present members of staff could keep in touch with one another and the College. General Morrison proposed that the Duntroon Society should be set up and run from the College with recognition of the need for a structure in each State or Territory and in New Zealand. The membership included former cadets, past and present members of staff, widows and others with an interest in the College. General Morrison stated in 1980 that he did "not want the Society to be, or be seen to be, simply an old boys' club".

The Society charged member subscriptions in order to be self-supporting and to publish a biannual newsletter. He concluded his 1980 introduction by indicating that the Society "offers the means to continue the strong bond of association that was so much a part of our time at Duntroon".

I was one of the many who enthusiastically embraced the Society and enjoyed attending the Autumn Lunches and other functions at Duntroon House. Not least of the appeal for many to join the Society was the great admiration in which we held Alan Morrison.

On moving to Melbourne in 1982, I took part in activities in the Victorian Branch, but did not have any management responsibilities. Soon after moving back to the ACT in early 1985, I joined the ACT Branch committee. More on this to follow.

My involvement with St John Ambulance (the major work of the Venerable Order of St John) began in the late seventies at the suggestion of Dr John Alwyn, who was the Senior Medical Officer of the Defence Health Centre at Russell Offices. I had qualified at a St John First Aid Course when I was a Senior Scout in the early 1950s but did not have any other contact with the organisation until John Alwyn asked for my help with the administration of First Aid training in the ACT.

I became the Secretary/Treasurer of the ACT division of the NSW Branch and was mainly concerned with the First Aid teaching work of St John. The ACT Division had an office in the Priory building and conducted First Aid Courses in the hall. At the time I expected to provide some short-term help to Dr Alwyn, but, in fact, it was my introduction to what became an association with the Order that lasted over thirty years.

I was appointed a committee member of the ACT Branch on its formation and remained so until I moved to Victoria in 1982.

Before proceeding, I must stress that I do not intend this book to be a rigorous history, but simply a tale of what seemed important or interesting. There may be chronological errors and other inaccuracies because of my failing memory, but none will be deliberate.

JOINT COMMUNICATIONS – ELECTRONICS BRANCH
1985

I assumed the appointment of DGJCE on the 21st January 1985, just two weeks short of thirty years after I entered Duntroon. I now held the senior communications position in the Australian Defence Force, with an extensive range of responsibilities, both national and international.

A major change to command arrangements occurred in 1984 with the amendment of the Defence Act to provide for the Chief of Defence Force Staff becoming the Chief of the Defence Force with the attendant command responsibilities. There was no change to the diarchy of control, so the CDF and Secretary were still jointly responsible to the Minister. This change did not flow on to other appointments in Headquarters ADF so I was still just a Branch Head, only able to direct my staff and achieve outcomes across the Services by persuasion – not always easy.

At the time I arrived, the CDF was General Sir Phillip Bennett and the Secretary Sir William Cole. I had known Sir Phillip for many years and was more than happy being on his staff. I met Sir William soon after my arrival and found him to be courteous and approachable. The diarchy seemed to work harmoniously and effectively although there were still frictions between the civilian and military staffs. I had regular contact with both as the principal source of communications-electronics advice.

Since leaving the Branch in 1979, there was a change of title from Joint Communications Branch to Joint Communications-Electronics Branch and the branch organisation had expanded to include a Directorate of Strategic Communications.

The major tasks of the Branch at the national level were management of the strategic communications system, promoting interoperability of Single Service communications and electronic warfare, spectrum engineering and management, and developing engineering standards for communications – electronics. Most of the national work was conducted through the Defence Communications Committee (chaired by me) and a subordinate committee, The Defence Communications Coordinating Committee, chaired by the Director Policy and Plans on my staff.

The international work of the Branch centred on the Combined Communications Electronics Board (CCEB) established during World War II to ensure, not always successfully, communications interoperability between the Allies. The role of the CCEB had hardly changed since the war and many of our national responsibilities flowed directly from it. I was, ex officio, the Australian Principal on the Board. My DGJCE responsibilities also extended to the Joint Facilities that required me to get additional security clearances.

Another responsibility that came with the appointment was chairman of the Board of Management of the Defence Journal. I do not know why this appointment devolved to DGJCE, but it was probably because of someone making the connection between communications and a journal!

The opening days of my appointment were filled with courtesy calls and takeover briefings, and I was much assisted by my predecessor, Commodore Harry Adams. One of the first calls was to the Director General Operations. The conversation came around to command and control whereupon the DGOPS said to Harry, "What would you know about command and controls – you are a communicator". Harry nearly exploded as he had had several command appointments, including flotilla command. I mention this now, as this attitude was not uncommon and was raised with me frequently.

* * *

I had barely settled into my office when I went to New Zealand to attend a CCEB conference on spectrum management, as it affected interoperability. This was the first of many overseas trips as DGJC-E and, I must admit, that I enjoyed the travel demands of the position.

Back in Canberra, I became quickly immersed in the workings of the Branch through the various committees and briefing sessions. I had inherited a very competent staff from Harry Adams and, most of the time, I could sit back and observe with the occasional forays into the Defence Department to argue the case for communications and electronic warfare projects, in particular, the High Frequency Radio Enhancement Study and discussions on satellite communications.

As mentioned above, I had ex officio responsibilities for some aspects of the Joint Facilities, Nurrungar, North-West Cape and Pine Gap. These responsibilities required that I have some particular positive vetting security clearances. One referee I nominated, my classmate Steve Hart, told me years later that he was grilled for about six hours on my background and he further stated he was amazed at the depth of their knowledge of me. As part of the process, I was interviewed by a very young American who asked me if I gambled. What followed caused me some amusement:

"Do you gamble?"

"Yes."

"On horses?"

"No."

"On dogs?"

"Certainly not!"

"Then what?"

"Poker Machines, often with my wife."

"Do you gamble a lot of money?"

"Well, that's relative – we have no dependent children."

At this stage he gave up and soon after I was advised that there was no impediment to my fulfilling my responsibilities.

Over the course of my appointment, I made several visits to these establishments and was very much impressed with the facilities themselves and the professionalism and dedication of the staffs.

In June 1995, it was back to New Zealand for a full meeting of the CCEB during which I was presented with my CCEB tie – a most important event! Much of the meeting was taken up with standard issues affecting interoperability and detailed briefings of communications-electronics projects that were of current or potential interest. Most items on the agenda were classified and so there is little I can add here.

Soon after returning from New Zealand, I had a visit to Adelaide to attend my first national meeting of St John Ambulance. This was my first acquaintance at the national level and I was much impressed by the keenness and dedication of the members of its various branches. The Deputy Prior of the Order in South Australia, the State Governor Lieutenant General Sir Donald Dunstan, held a reception during the meeting and I had the great pleasure of a brief conversation with him harking back to when he was Company Commander of Alamein Company at Duntroon.

It was about this time I started meeting with John Sewell, who was the First Assistant Secretary Computer Services Division. While at EDE, I had become increasingly interested in the concepts of information management and I soon found that John was similarly interested. There was a clear divide between administrative and operational computing systems that was not particularly productive, as much operational information was held on administrative systems. There was great distrust on both sides fuelled by 'patch protection'. I will return to this topic later.

In August 1985, I was invited by AVM Collings to accompany him on a visit to Butterworth so that I could observe the work of the P3 Orion aircraft patrolling the Straits of Malacca. I accepted with alacrity, and set off for an enjoyable and interesting week away from Canberra's winter.

My first sortie on the P3 was exciting. My pre-flight briefing was minimal, and I settled in for a leisurely flight to get 'on-station'.

I was looking out the window of the P3 at the start of the observation run when suddenly one engine shut down. I reported this to the flight deck. My concern caused great amusement as the captain explained this was standard operating procedure to allow a longer time loitering on station. As we continued on our track to Sri Lanka, we came across a Russian submarine making a surface transit. What followed was exciting indeed as we made several passes at a low enough altitude for me to observe Russian submariners shaking their fists at us!

Once we were again on our way, the captain asked me if I would like a 'drive'. I settled into the right-hand seat and the captain instructed me on the various controls. They gave me control and then I received such commands as "right hand down", "left hand down", "straighten up", etc which led me to believe this piloting lark was a piece of cake!

On reaching the eastern side of Sri Lanka, we reversed course and started the return flight. As we approached Butterworth, the captain asked if I could provide some help. I readily agreed and was then surprised to find that my task was to sit on the window ledge of the cockpit and keep my eyes peeled in the right-hand quadrant looking for aircraft. They assured me this was a most necessary task, as the crew had little faith in the local air traffic control.

During the next two days waiting for the next sortie, I was honoured by the RAR Company in Butterworth with a Quarter Guard and did some sightseeing and try the local tailor who made me a suitable open-neck shirt to attend a dinner for the DCAFS. There was some excitement during the dinner when a couple of young fighter pilots let off some fireworks, an act specifically forbidden by the CO, which resulted in some swift justice. There was no permanent career damage as I learnt recently that both made star rank!

We had an uneventful trip home and, luckily, only a cursory inspection by Customs at Darwin. We had a tour of RAAF facilities in Darwin and the following day to Tindale for a base inspection before continuing to Canberra. All this fun and being paid for it!

No sooner back in Canberra and I had another adventure, this being an overnight trip on an 'O' Class submarine. I went to HMAS *Penguin*, changed into 'greens' and embarked to observe an exercise which included firing inert torpedoes.

I'm not sure what I expected when I boarded the boat, but it was certainly not the reality. It could have been a scene from *Das Boot* – the only reason I recognised the Captain (unfortunately, I cannot remember his name) was because they introduced me to him. The dress was as informal as it could be, with no sign of badges of rank or other distinguishing marks denoting authority. This was, of course, completely understandable because of the size of the crew. What struck me immediately was the over-riding atmosphere of competence, efficiency and enthusiasm.

After a short cruise on Sydney harbour, the boat submerged to join some surface ships around the Nowra region. After submerging, I found the sensation quite eerie as there was hardly any discernible movement of the boat. I was able to move freely about and spoke to a number of the crew about the submarine service. Most were disdainful of the surface Navy and had no desire to trade their lot of living in confined spaces to being on the surface "with your head in a bucket" most of the time. Much later in the day I experienced close living when I tried for a few hours sleep in the Wardroom. My bunk left me with only inches between my nose and the bottom of the upper bunk!

The exercise involved firing practice Mark 48 torpedos. In one case there was a problem with the wire guidance system with part of the control cable being caught in the torpedo tube door (this explanation is, no doubt, inadequate and possible inaccurate, but it is my understanding as to the fault situation). I was invited down to the torpedo room to observe the fix. It was an interesting situation with a hoary old chief working on the problem with a young Acting Sub-Lieutenant offering policy guidance. The Chief was most tolerant of this until the young officer suggested the Chief crawl down the tube and recover the wire. Thereupon, the Chief's head emerged from the tube and offered the suggested "Why don't you go back to the Wardroom and have a cup of tea and a good lie down". The young officer being well trained replied "Good idea Chief" and hastily fled the scene. I was immediately reminded of my experience with WO Joyce Curran as detailed in my earlier book.

A problem arose with a 'lost' torpedo which meant that we needed to stay on station during the search so that our return to HMAS *Penguin* would be very much delayed. The Captain appraised me of this and said that we would not be back in time for me to catch my return flight to Canberra. I assured him this was of no consequence and I continued my conversations with members of the crew. About half an hour later the Captain found me and said the problem was solved as an FFG involved in the exercise was sending its helicopter to transfer me to the ship and so return me to Sydney in good time.

Now this is something I hadn't bargained on. The thought of being winched from the boat up to the helicopter was verging on the terrifying but I felt unable to reject this 'kind' offer. In any case it was too late as the helicopter was already airborne and close by. I was briefed on the procedure to be followed. The weather was too rough for winching from the deck so the recovery would be from the conning tower. I was invited to have a look at the approaching helicopter through the periscope and was rather appalled by the rough sea. While I watched I listened to an exchange between the submarine and helicopter as to who would provide the necessary lifejacket.

The time came, and I was assisted up the ladder to the top of the conning tower and watched as the helicopter lowered a sling. Another hoary old Chief said that once I had my arms through the sling, I should extend them horizontally to signal I was ready for the lift. I politely advised him there was no way that I would relinquish my hold on the cables. I should add that the lifejacket issue had not been resolved, so I left the submarine without one. I was winched up at a remarkable rate and got caught up in the skid of the helicopter but was quickly brought aboard by the copilot who then handed me a lifejacket while informing me it was an offence to fly over water without one!

After a short journey we landed on the rear deck of the FFG where I was received with some ceremony and escorted to the bridge where the Captain rather pointedly offered me the facilities of his cabin to refresh myself. I learnt that it did not take long for the smell of diesel to penetrate deeply into the pores of the skin.

So I got back to Sydney in time to catch my flight home to Canberra and as I said before "all this fun and being paid for it".

About mid-year AVM Billie Collings became Deputy Chief of the Air Staff handing over to Major General Neville Smethurst (Class of 56). I had enjoyed working with the AVM and looked forward to working with his replacement.

I received an invitation to take part in the Armed Forces Communications Electronics Association (AFCEA) Convention in

Seoul, South Korea. With little expectation of receiving approval, I approached ACOPS, who readily agreed to my going and also agreed that I could extend my visit to include Hong Kong.

What an experience! Whenever the conferees moved about from venue to venue, we had a Military Police escort with flashing lights and the whole nine yards. The conference went well, and I was able to re-establish connections I had made with members of the USA Signal Corps and civilians I had worked with at SATCOMA and MALLARD. There was a fair representation there from the UK and I saw old friends from Plessey and also caught up with my classmate, Steve Hart. The conference was of much more benefit than I had expected and it renewed my interest in AFCEA.

The side trip to Hong Kong allowed me to visit a UK facility located close to the border with China. It was interesting to look through binoculars to observe interested people looking back at us! The visit was useful and I will leave it at that.

On my last night in Hong Kong, the RAAF attaché (whose name escapes me) invited me to have dinner at a local restaurant specialising in Hot Sour Soup. As this was a favourite dish of mine, I readily accepted the invitation with the proviso that it would be my treat. My host explained that, because of the spiciness of the soup, this would be the last course. At the appointed time, the waiter brought the soup in a large tureen and presented it to my host, who immediately declared it was not right and instructed him to fix it. The waiter retired and returned almost immediately, having 'fixed' the soup by pouring chilli sauce around the surface. While returning to our table, he apparently commented to the other patrons that "round eyes did not know about Chinese cooking" obviously unaware that my host was fluent in Mandarin and Cantonese. My host then berated our waiter and his ancestors in both languages, to the amusement of the other patrons and the very visible 'loss of face' of our waiter.

* * *

Back to Canberra and the routine of the Headquarters. This was not hum drum but an unrelenting round of conferences, briefings, committee meetings, meetings with industry representatives, visits to Service and research establishments, etc while being very much involved in the development of policy papers and contingency plans. I found the appointment intellectually stimulating and professionally rewarding.

I rounded off October 1985 with a week Acting Assistant Chief of the Defence Force-Operations, which involved attending the CDF and Secretary's meetings and briefings. This was interesting, to say the least.

There was not a great deal of note in the last two months of the year apart from the usual round of end-of-year functions, farewells, etc. I had had a good year and was very comfortable in my place at the higher echelons of Defence.

Before starting this section of the book, I reviewed my diaries for the year. I soon discovered that the diaries were of little use as most entries comprised a time and a person's name or an abbreviation/ acronym. Most of the names now mean little to me, but I expect they were mostly representatives of the communications industry. Among the names were Brigadier Ray Sunderland, who represented British Aerospace and who now is my regular golf companion. There is also evidence of regular meetings with representatives of the Australian Satellite Communications Agency (AUSSAT), the Defence Signals Directorate (DSD), the Defence Science and Technology organisation (DSTO), Foreign Affairs communications and Telecomm.

I must admit to memory lapses regarding some acronyms and reference to Dr Google was of little help. As a case in point, I have an entry MITSG for 09:30, 28 February 1986 and at various other times throughout the year. I could not locate the meaning of this acronym and so consulted Stuart Althaus, who had been one of my SO3s. He came up with:

"Maritime International Training Support Group (MITSG). The MITSG is a 'working group' made up of friendly governments focused on effective and efficient military training that incorporates proven training methods and technologies. Its primary goal is to promote collaborative working relationships that will result in the development of effective training technologies based on sound human performance principles. All participants collectively discuss and exchange information on emerging technologies and their application within training and performance support."

This did not help much, as I could not understand why I would have been involved with such a group. Stuart is doing further research on this topic.

So now I will again have to rely on my defective memory. So I must repeat the disclaimer: I apologise for errors and omissions while I do my best to provide a reasonably accurate account of the JCEB in 1986.

JINDALEE

In late February 1986, I attended a brief to the CDF on the Jindalee Operational Radar Network (JORN). I do not recall previous meetings, but there were a number to follow as approval was being sought by DSTO for the design and development of the network.

The Weapons Research Establishment, part of the DSTO, had been researching the ionosphere from the 1950s. From 1970, the Jindalee Over The Horizon Radar (OTHR) became a core research project to provide surveillance across Australia's northern air and sea approaches.

The first experimental radar was built in the mid-1970s near Alice Springs and became known as 'Stage A'. 'A' was modestly powered and had a narrow field of detection, but it detected aircraft at long range and ships.

DSTO developed Jindalee 'B' in the early 80s at a cost of approximately $M30. 'B' was higher powered and had a wider scan angle. It also had radar scan-while-track and advanced frequency management.

I must admit that I was not a supporter of the project because of its reliance on the highly unpredictable ionosphere and the possible deleterious effects that its high power, high frequency radiation would have on communication circuits. Another factor was that the USA had abandoned this technology after considerable expenditure. However, Jindalee enjoyed a high level of support within Defence and Government and in October 1986, approval was given for the design and development of the OTHR network. The winning contractors were Telstra and GEC-Marconi with a completion date set in June 1997.

CCEB 1986

The annual Principals Meeting was to be held in the USA in June 1986. The meeting itself was only four days duration so, as was the usual practice in such cases, I organised side visits on the West Coast, the AFCEA Convention in Washington, relevant organisations in Fort Monmouth, and a visit to Canada. As Australia was to host CCEB 1987, I sought approval for Captain Stuart Althaus, who would be responsible for administration for the conference, to accompany me. Stuart kept a journal of the trip, which has been a great help in writing this section. This trip was very much an adventure for Stuart and, to give him his due, he soaked up every minute.

We flew out of Sydney on 24 May on what should have been a direct flight to San Francisco, but they diverted us to Brisbane to collect passengers stranded when an United flight had a breakdown. The upshot of this was that we landed in San Francisco exhausted after a long trip. After dropping Stuart and our luggage at our hotel, I found a place to park and then promptly forgot the location. As a result, we spent some time on our first day in the USA looking for our car that we finally found carefully parked in a lane. The episode was a source of amusement to Stuart, as I had given him a ribbing about allowing our briefcases to be boarded as luggage in Sydney. In his words: "we are even".

We used our rest day for sightseeing and the following day undertook the cross-country flight to Washington, DC. We were met by our Escort Officer, Squadron Leader Greg Hockings, from the Embassy. This was my first overseas visit that warranted an escort, no doubt because of the importance of the CCEB.

Our first working day was a visit to the National Security Agency at Fort Meade. They made us most welcome, and it was a thoroughly enjoyable and informative visit.

We visited Andrews Air Force Base and received a briefing on Mystic Star, the communications system for Air Force 1, as a prelude to attending the AFCEA Convention.

The AFCEA Convention lasted for the rest of the week and included official luncheons and dinners plus an absolutely extraordinary display of communications-electrons systems and equipments. Apart from seeing the latest in technology, the Convention provided an ideal opportunity to make and renew contacts with both the military, public service and industry.

Another free weekend, so I dragged Stuart to Williamsburg on the Saturday and the Smithsonian on Sunday. It so happened that Deryck Scarr, my eldest step-daughter's husband, was in Washington, so he came along with us.

Started the new week with a visit to the Director of Army C3, MG Donahue, at the Pentagon and then to HADS, Washington, Major General John Coates, who later became Chief of Army. In the afternoon we visited LGEN Powers of the Defence Communications Agency, followed by a courtesy call to LGEN McKnight, who was to chair the CCEB.

The CCEB started the next day with each of the Principals addressing the gathering. I cannot remember what I said, but Stuart assured me it was well received. All the Principals were concerned with Single Service decisions that had the potential to weaken interoperability across the board. Our discussions led to a directive to overcome this problem.

That evening Colonel, later BGEN, Don Banks, the Canadian Principal, and I engaged in a marathon exchange of jokes. This became a feature of all our subsequent meetings and cemented the long-standing friendship we shared.

The CCEB meeting went well and resulted in several decisions that would have long-standing effects. One particularly pleasing result of the meeting was that the US and NZ agreed NZ should remain on the board notwithstanding difficulties with the ANZUS Treaty.

The meeting concluded on Thursday 5 June, with the gavel being passed to me to chair P17M scheduled for 1987 at Watsonia Barracks.

Now with Washington behind us, we drove to Fort Monmouth in New Jersey, where I had spent two-and-a-half years in the 60s.

The drive up from Washington was uneventful, with Stuart sleeping most of the way. It did not surprise me as he had led a fearful pace of work and play since arriving from Australia. The drive took the best part of the day and we just had time to settle in at our motel before going to cocktails hosted by an old friend, Lieutenant Colonel Tom Davies and his wife, Pat.

Sunday was free so we all, Stuart, Greg and I, drove up to New York city for the day. Stuart was unimpressed and described the city in such terms as 'real dump' and 'filth' hole. He did, however, enjoy the service at the Central Presbyterian Church while I went to Mass at St Patrick's Cathedral.

Back to work on Monday with visits to the Joint Tactical Command, Control and Communications Agency with briefings on various interoperability issues. The Director of JTC3A was MGEN Norman Archibald, whom I had known when we were both Majors at Fort Monmouth. We spent the afternoon at the Satellite Communication Agency, which was quite a nostalgic trip for me.

We had an easy morning on Tuesday at Communications-Electronics Command with Tom Davies and then to La Guardia Airport for a flight to Ottawa.

Don Banks, our host in Ottawa, took us to the recently renovated Army Officers Club. I gather that the renovation started with internal painting that required some plaster sheets to be replaced. Removal of the sheets revealed beautiful wood panelling, so restoration became a major project costing close to a million dollars.

We flew to Quebec after a visit to the National Defence

Headquarters and booked into the Frontenac, an old Canadian Railways property. My old friend Jim McKeown from my Monmouth days in the 60s had come up from New Hampshire to meet us. Stuart was mightily impressed with Jim who, when told about the state of his room, promptly engaged with the management and that resulted in Stuart being moved to the 11th floor with a complementary box of chocolates.

Jim and I had a great time reminiscing about our adventures in Monmouth. Jim came from a wealthy family but enjoyed being a soldier. When he retired from the Army, he bought a bank!

Jim was a regular visitor to Quebec. He took us all to an extremely nice restaurant, the Café de Paix, for what I can only describe as a wonderful meal. Stuart again showed his youthful stamina by going clubbing after leaving the rest of us at the Frontenac at 01:30.

My reason for detouring to Quebec was to allow me to visit the Defence Research Establishment at Valcartier. The establishment was very similar to the Engineering Development Establishment, but with more emphasis on research.

The visit did not go well. They viewed our visit as an unwarranted intrusion into their work. The attentions of the officious female escort did not help the atmosphere. To quote Stuart: "She won the pain of the day award against a stiff field."

At the end of an uninspiring day, particularly when compared with the rest of our visit, we flew to Vancouver for a few days en route to Hawaii. Bid farewell to Greg Hockings, who had been a great help throughout the visit.

* * *

We arrived in Hawaii late on the evening of Saturday 14th June to be met by MGEN Bob Lynn, an USAF Protocol Officer and an escort officer whose given name was Kent but whose surname escapes me.

Two full working days with briefings and visits including Commander-in-Chief Pacific Command, the National Military

Command Centre, the Defence Communications Agency Pacific and the Naval Communications Command East Pacific. A very full programme interspersed with memorable hospitality.

Departed Honolulu around midnight to arrive back in Australia early Thursday morning after a hectic five weeks. This was a most successful trip, with several significant achievements in interoperability and arrangements for co-operation.

After the long trip to the USA and Canada, I was looking forward to five days leave, but I had to forgo this to act in the ACOPS role while he was away. This was a regular occurrence that I found extremely interesting and enjoyable.

My next commitment was the AUS-US Comms forum to be held in Hawaii in early November. On the off-chance I applied to take some leave after the meeting. To my surprise this was granted so I took Shirley with me. We stayed on base for the meeting and then moved to the Hale-Koa, a military hotel resort, for my leave.

The Hale-Koa was one of many resorts run by the US Army. Like our Australian Army resorts, rates were set by rank to make the resort affordable for all. I expect that the booking was organised by General Lynn so Shirley and I had an upper floor suite with a reserved parking space. This was extremely convenient as we did a lot of day trips around the island and didn't have to worry about finding a space on our return. The parking space had my name on it with a sign:

"Don't even think about parking here".

The Hale-Koa had a number of dining rooms from the formal down to a beach-side burger bar. Shirley and I were on the beach one day and she wanted a bread roll for some reason. The burger bar attendant said he couldn't supply this by itself as they were for hot dogs. So I then asked for a hot dog but said he could keep the frankfurter. This, he said, was not possible as it would upset the stocktake!

On the Sunday during our stay, Shirley went to the beach while I went to Mass. There was an outdoor church service in progress and as she drew near, the leader welcomed her to attend and made a fuss over her being an Australian.

All in all we had a most enjoyable holiday on top of a successful forum.

I spent much of the rest of the year involved with several command, control and communications projects from the interoperability aspect. These included RAVEN, combat net radio; PARAKEET, area communications including access for the RAAF; AUSTACCS, the command-and-control system; and the High Frequency Radio Enhancement Project. I also had some involvement in JINDALEE and the Australian and USA Joint Projects.

THE PRINCESS ANNE BANNER

Approval for the Corps to be honoured with the Process Anne Banner was obtained in 1980 while I was Director of Communications. This was followed by a process involving draft designs with approval by the Princess at every stage.

The ultimate design depicted the Commonwealth Coat of Arms on the Obverse and the Corps Badge centred on the Reverse with the Colonel-in-Chief's Cipher, A, placed at the top of the Banner on the staff side, all with a red background.

Throughout the design phase, action was taken by the Corps Committee to raise funds for the necessary accoutrements for the Banner escort and a suitable container for shipping. The latter item was important, as the Banner would be made available to various units for major parades and celebrations.

I cannot remember the total costs, but they were substantial.

The design, approval and manufacture process took some five years, so the Corps were ready to receive the Banner in 1986. The obvious choice was Corps Day, the 10th November, but the parade was deferred to the 29th to accommodate the Governor-General's programme.

I was looking forward to the festivities and my wife, Shirley, and I were at Canberra Airport along with Lieutenant General Peter Gration, Chief of Army, in the early AM for a flight to Melbourne. We were close to boarding when Murphy's Law struck with a vengeance. A light aircraft had made a poor landing and was blocking one of the taxi-ways. The upshot was that Ansett could fly, but Qantas could not, so the General could get started, but Shirley and I were delayed. The delay became lengthy, and we had to abandon our plans to attend, which was a major disappointment.

However, the Banner Parade was a great success and was the subject of a magnificent painting.

* * *

The year drew to a close with the usual round of end-of-year functions, farewells, etc. as the posting cycle took its usual course.

I was well satisfied with the Branch's performance during the year. I had a well-integrated team that worked well together and was not afraid to express their individual opinions. What more could I expect?

INFORMATION SYSTEMS

I had intended to include my involvement with computing systems into the overall tale of the Branch but this became more difficult as the writing progressed – hence the separation. While not specifically within my range of responsibilities, I had become interested in some large computing projects that were extant when I arrived. I cannot remember how I got myself invited to the various project meetings but I expect it was through the good offices of John Sewell who headed the Computing Services Division.

Before proceeding, I would like to review the general state of computing in the wider community to give my comments context.

The history of computing is well known so I will step back only to the 1950s. Grace Hopper developed the first computer language that became known as COBOL. FORTAN that was developed by a team at IBM in 1954 closely followed this.

In 1958, just eleven years after the transistor was invented, Jack Kilby and Robert Noyce unveiled the first integrated circuit that became known as the computer chip.

In 1964 Douglas Engelbart unveiled a prototype of a modern computer with a mouse and a graphical user interface. Thus a specialised machine for scientists and mathematicians evolved to a technology more accessible to the general community.

In 1969 a group from Bell Labs produced the operating system UNIX that addressed compatibility issues that allowed its use over a range of platforms. UNIX remained as a mainframe operating system for many years but eventually became the basis of personal computer operating systems.

In 1970 the first Dynamic Access Memory chip was released by Intel. In 1971 IBM developed the 'floppy disc' allowing data to be shared among computers quickly followed in 1973 by the development of Ethernet.

Between 1974 and 1977 several personal computers from IBM, Radio Shack and Commodore appeared on the market. In 1975 Paul Allen and Bill Gates wrote software for the Altair 8080 that was so successful that they formed Microsoft.

In 1976 Steve Jobs and Steve Wozniak started Apple Computers and produced Apple 1 with a single circuit board. A year later Apple II featuring colour graphics and an audio cassette for storage was shown at the West Coast Computer Fair.

The first computerised spreadsheet programme, VisiCalc, was introduced in1978 followed a year later by the word-processing programme, WordStar.

IBM introduced its first personal computer, ACORN, in 1981 featuring Microsoft's MS-DOS, two floppy disc ports and an optional colour monitor.

In 1983, Apple released Lisa. This machine had a GUI and featured drop-down menus and icons. It was not a success but it evolved into the Macintosh and led to the first portable computer marketed as a 'laptop'.

In 1985 Microsoft announced Windows and Commodore released the Amiga 1000 that featured advanced audio and video capabilities.

We are now up-to-date and the story can proceed, hopefully without resorting to 20–20 hindsight.

Computing was relatively new in the Department and the ADF but there was already a line drawn between 'operational' and 'administrative' computing. 'Operational' computing included command and control, weapons, navigational and communications systems. The 'administrative' systems included the personnel, financial and logistics functions. Desktop computing was limited, and it was apparent that there was considerable uncoordinated effort being applied at the unit level for a variety of administrative uses.

The distinction between operational and administrative computing was flawed in that much of the information on readiness (personnel and equipment) was held on administrative systems and was not readily accessible to planners.

I was aware that many Army units had used regimental (i.e. non-government) funds to purchase desktop machines that were used by programming enthusiasts to develop systems to assist such administrative functions as quartermaster accounting, leave schedules, attendance records in reserve units, etc. I suspect that similar activities were occurring in Naval and Air Force establishments.

Within Army there was considerable discussion on what Corps should be responsible for computing in general. Ordnance made a claim on the basis of its long term association with store base computing systems while Signals held that its involvement in command and control systems made it better qualified for operational systems. For my part, I was attracted to the US Army arrangement whereby the US Signal Corps was responsible for information in all its forms including libraries, printing, technical specifications, video production, etc. I didn't get far with promoting such an expansion of the responsibilities of RASigs but I still considered that the concepts of information technology should be pursued.

Defence was not the only large organisation coming to grips with large computing projects. The literature of the time was filled with horror stories of time and cost overruns and even abandonment of entire projects.

There was also considerable argument about how to approach large computing projects. One of the major problems was the lack of experience with the possibilities of computer based solutions to a myriad of problems. Not helping was the reluctance of senior people to state their requirements. I had personal experience of approaching senior military officers in both command and staff appointments with questions on their needs and being fobbed-off to their IT advisers who were invariably more interested in such technical solutions as operating systems, etc. I think it would be fair to say there was widespread suspicion of computing systems.

The Supply Systems Redevelopment Programme (SSRP) was indicative of a number of large IT projects that were underway during my tenure as DGJCE. As I remember it, SSRP was to select an existing solution that was the closest match to the Statement of Requirement and then modify the code as necessary. On the face of it this was a sound approach but the project was beset with major problems and, in the end, did not produce much more that a minor application suitable for 'Q' Store management. Many millions of dollars were spent and the project was way over time when it was eventually closed.

Another ill-fated project was one for personnel management. This started well with a major effort to determine the user requirements. However, it later emerged that the 'users' that were consulted were at the lower echelons and led to some very dubious outcomes.

I do not wish to comment on the causes of these project failures except to say that it was not unique as there were many similar stories in other government departments and civilian enterprises. Perhaps many of the failures could have been caused by the requirements being too ambitious given the state of development of computer systems at the time.

I did continue to press the case for combining communications and computing into informations systems with the support of the previously mentioned John Sewell. There was little support for us at the time notwithstanding the rapid take-up of the concepts by industry and some allies. All was not in vain because many years later a Chief of Information was established in Defence.

As I entered my third year as DGJCE, I gave some thought to the next step in my career. I had my replacement, Jerry Tipping, well and truly lined up and was more than confident that I would leave the Branch in good hands.

1987 was to be a year of great change as General Sir Phillip Bennett was due to retire in April with his nominated replacement being General Peter Gration.

I had very much enjoyed working with General Bennett. I found him most approachable and affable in addition to being very good at his job. I had seen him in action on a number of occasions in the Chiefs of Staff Committee meetings when I acted as Assistant Chief of Operations. He invariably was able to achieve unanimity among the Chiefs but he also was quite willing and able to assert his authority when necessary.

The British invited me to attend the 1987 Electronic Warfare Inwards Mission in London from the 4–18th April. As the organisers were paying all expenses, I did not have to concern myself with gaining the usual travel approvals.

An old friend, Graeme Burgess, was Defence Attaché at the time, and I could share some time with him and Sally.

The Inwards Mission programme was very busy and included several site visits to government laboratories besides the leading commercial suppliers of electronic warfare equipment.

There were attendees from most of the NATO countries and a fair collection from the middle east. They provided us with accommodation in a top class hotel and covered our laundry and mini-bar expenses. Not covered was international telephone calls. Although it was clarified that such calls were to be a personal expense, several attendees ignored the warnings and this led to some tension at the end of the Mission.

I took advantage of the weekend break to hire a car and drive to Wales, much to the amazement of my British hosts, who thought the round trip of some 650 km would be suitable for a week's holiday! I very much enjoyed the time off and started the second week refreshed and ready to go.

I found this visit to be rewarding, not only for access to the latest in UK EW systems but also for the contacts made with industry.

CCEB PRINCIPALS MEETING
JUNE 1987

The major tasks of the Joint Communications-Electronics Branch were advice to the CDF and Secretary on interoperability issues at national and international level. At the national level, there were inter-service committees that sought to ensure interoperability at equipment and doctrine level between our own Services. At the international level, there were the Single Service Multilateral Meetings with the final responsibility being with the CCEB.

There was a great deal of preparatory work done in the Branch in the build-up to the meeting which I was to chair. The meeting was to be held at the School of Signals in Watsonia in the conference facilities at Meares House.

All delegates and their staffs were to be accommodated at the Park Royal in Melbourne for ease of administration and also to allow informal contacts at the various levels.

My party took up residence on Saturday the 20th so that we could greet the delegates who were all due to arrive on the following day.

The American Principal was Admiral Jerry Tuttle. The Admiral, whom I had not met before, was a fascinating figure. He was an iconic, even cult, figure in the US military and many allied militaries and is generally credited with being the inventor of modern command-and-control systems.

Despite all this, he was unassuming and friendly, and we struck up a friendship on the car ride from the airport to the hotel. I think it was on this trip that he asked me if I had a nickname. I assured him I had more than one – Tubby Evans from the day I entered Duntroon and Bionic Budgie after my first visit to the Corps Sergeants' Mess as Director of Communications. He told me he was known as SLUF, which he pronounced more like slurf. He explained that this translated to Short Little Ugly F...er, a name given because of the facial damage received from ejecting twice from his fighter plane at high speed!

The British delegation was led by Major General Gordon Oehlers, who had also attended P16M in Washington. This visit was to be his last before retiring from the Army. He was a most interesting man and had reached high rank by sheer talent and against the odds. Despite others perceiving him as 'the odd man out,' he achieved the rank of General. In Washington, I had assured him I would arrange his 'Dining-Out' at Watsonia and all was in hand for this to happen. I discovered later that they appointed him one of the Colonels Commandant of Royal Signals almost immediately on his retirement.

Colonel Don Banks was the Canadian delegate and I regret to say that I have forgotten who represented New Zealand.

The culmination of the CCEB was the formal dinner held in the Corps Mess at Watsonia. Admiral Tuttle was to be the Guest of Honour and I accompanied him from the city, planning to arrive at about 15 minutes before dinner. Murphy's Law went into top gear and we arrived at Watsonia far too early, so we drove around the block. Then we ran into a delay, so when we arrived at the Mess the bugler was playing 'Fives' – a five-minute warning before dinner. The Admiral jumped to attention and saluted, thinking this was a bugle call salute. He said to me, "you Aussies sure know how to do things right" and I didn't have the heart to correct him!

The attendees suitably farewelled Gordon Oehlers and he made a most gracious speech, as did the Admiral. I took this opportunity to hand over the gavel for P18M.

The meeting went extremely well with much of the credit due to my staff, who all worked hard.

1987 had been a busy year with my involvement in a number of major communications-electronics projects and with increased involvement with the Joint Facilities (Pine Gap, Nurrungar and North-West Cape). All in all, a busy but professionally rewarding and enjoyable year was to end earlier than I had expected.

I had been very vocal about how much I had enjoyed EDE and had often remarked that I would return in a flash. Unexpectedly, they took me at my word. George Salmon, who I had handed over to in 1985, had lost a son in a tragic motorbike accident and had requested a return to Canberra. At about this time, Ian Meibusch resigned, with George being the obvious choice for his replacement. So in 1987, they posted me back to EDE.

I had an interesting exit interview with General Gration, who was surprised that I had readily accepted a return to EDE. I assured him I was happy to go as I did not expect a long posting.

Rather than intersperse my extra curricula activities with my memories of DGJC-E, I gathered all such happenings in a separate section.

Harry, the late husband of my wife, had been an active Rotarian and Shirley had continued her membership in Inner Wheel after his death. I had shown an interest in Rotary and soon after our return from Melbourne, I was invited to be interviewed as a prospective member of the Canberra-Woden Rotary Club. They duly elected me, and so I started a twenty plus year involvement. When we returned to Melbourne, I did not transfer membership to a local club but kept up my contacts at a distance and visited the Club whenever I was in Canberra.

* * *

I had remained active in the Duntroon Society while in Melbourne and resumed my activities immediately on my return to Canberra and became a member of the ACT Branch Committee.

By 1985, the Society had been in operation for only five years. The membership included many serving officers and retired alumni. The Society was in a state of evolution and there was much discussion about membership criteria, etc. I think at one stage there were six categories of member with different rights and responsibilities. The Society was self supporting, with members paying an annual membership fee of $25 with discounted rates for longer periods. Finances were not a problem as Newsletters were produced internally and the College paid postage.

There was usually only one social function a year in the ACT, this being the Autumn Lunch held in Duntroon House. During this period, attendance at this lunch was largely by serving officers.

As a Branch Committee member, I took part in national meetings. We conducted these in Canberra with State Branches being represented by 'proxies'. This system worked well and is still in use.

* * *

As with the Duntroon Society, I resumed my involvement with St John Ambulance in early 1985 and re-joined the ACT Committee. During 85–87, our preoccupation was raising money to purchase our own building. After much experimentation, we settled on car raffles as our major effort. This worked well for several years and, again after experimentation, settled on $1 tickets and multiple prizes. We found that adding first aid kits and courses made the raffle more attractive.

I found selling raffle tickets very interesting most of the time but dead boring at others. If I had a dollar for every time someone said: "you have to be in it to win it" we would have raised enough to build in record time. I also found it fascinating that many people insisted on recording exact directions to deliver their prize.

One shortcoming of the committee, as it existed in those years, was that membership seemed to be determined by an aspirant's willingness to sell raffle tickets. Initially, this did not present a problem but as our activities developed to include a major drive to expand First Aid Training and to start retail and wholesale selling of First Aid Kits, it became apparent that the Committee lacked several management and governance skills. In time, this problem was overcome, but we had some 'hairy' moments.

I had filled several roles on the Committee and the Chairmen asked me if I would consider becoming the Deputy Commissioner of the uniformed Operations Branch. I was happy to oblige, but my posting back to Melbourne overtook the plan.

EDE AGAIN

I took over command of EDE from George Salmon on 19 September 1987. I took up residence at the Naval and Military Club while I looked for suitable accommodation. Shirley was quite happy about moving back to Melbourne as her middle daughter, Deidre, was living there. Shirley was keen that we not live in the town house that I had purchased during the previous posting as it was rather cramped and unsuitable for entertaining.

I managed to locate a property on Brunswick Road that had been owed by an AFL team coach. The house was mock Tudor and had a very large swimming pool in the extensive grounds. What I thought would be a useful facility to have on hand, it became more of a nightmare and cost me a small fortune keeping the pool algae-free.

The house had another advantage in that there was a shed in the backyard which I was able to equip with tools to pursue my woodworking hobby. I never achieved a great level of skill but certainly enjoyed making sawdust!

Shirley took up a position with DVA in the city which was convenient for us both. We both settled in quickly and enjoyed frequent contact with Deidre.

* * *

I was warmly greeted by the staff at EDE but there was some hesitation and I found afterwards that there was some disquiet about who would raise the issue of my pipe smoking in my office (apparently a considerable effort was required to clean curtains, etc to remove tobacco odours). All were much relieved when I announced that I had given up smoking as a 50th birthday present to myself.

The transition back into EDE was painless and I was soon back into the routine of Division Heads meetings, project reviews and conferences, liaison with the various DSTO Laboratories, visits to industry, etc As in my previous posting to EDE, I was keen that our test facilities were known to be a Defence asset, so I also had regular contact with the other Services.

In October 1987, we had the Project Raven Review. This project was becoming problematic as Plessey were asking for an increasing number of specification relaxations that Army was reluctant to give.

Closeness to Watsonia Barracks also allowed me to keep in contact with serving and retired members of RASigs and Shirley, Deidre and I regularly attended Mess functions there.

As was the usual practice, EDE closed down over the Christmas period and so 1987 came to a satisfactory end.

1988

Early in the year, there was a AUS-US Communications Forum held in Watsonia and I was able to host General Lynn and his staff to EDE. It was a very real pleasure to show some hospitality and to show off the establishment.

Later in February, I was able to host the Chief of Army to a visit. I was very keen, as during my previous time with EDE, to raise the profile of the establishment and to demonstrate its value as a resource to enable the Army to be an intelligent purchaser of equipment.

In April, I visited the Small Arms Factory in Lithgow as some problems had arisen in the manufacture of the Steyr rifle. The problem was not that unusual and was a result of manufacturing changes not being properly recorded in specifications and drawings.

I had often felt that the facilities at Monegeetta were under-utilised so I approached Major General Cooke and offered him the use of the accommodation and dining facilities for 2 Division training purposes. This was accepted with some alacrity due to the relatively short distance from Melbourne.

In mid-April, I attended State Government House in Melbourne to receive the decoration of a Serving Brother of the Venerable Order of St John of Jerusalem. The award was in recognition of my work with St John Ambulance in the ACT. The medal was a silver eight pointed star usually referred to as the Maltese Cross.

I visited Canberra in May for the Presentation of New Colours for the Royal Military College, Duntroon. This was a most impressive event and it was an honour and pleasure to be present at it.

Shirley and I were back again later in the month to attend the marriage of my son, Damian, to Sue Paic. The wedding was at St Christopher's Cathedral and the celebrant was Bishop Morgan, Catholic Bishop to the ADF. 'Alo' Morgan was much loved and respected in the Army as he had been a chaplain during the New Guinea campaign in World War II as a young priest followed by years of dedicated to soldiers of all faiths. Unfortunately, this auspicious start to the marriage did not help and the marriage failed within the year.

* * *

My SO1 was Lt Col Peter Pridie, a member of the Armoured Corps and a tracked vehicle enthusiast as one would expect. He organised a visit to the Armoured Regiment and I had the opportunity to sit in the tank commanders seat while the Leopard tank's impressive mobility characteristics were demonstrated. Not only was I impressed, I was verging on the terrified as we drove at speed over all sorts of terrain!

Later in June, I was able to host a visit from the Industrial

Mobilisation Course this being part on my on-going programme to promote EDE.

* * *

In Armoured Corps circles the question 'wheels versus tracks' was the centre of endless and often heated debate. It came to the fore in the USA after the abortive attempt, Operation Eagle Claw, to rescue hostages held by Iran. Planners looked at possible alternatives to airborne missions and came up with wheeled armoured vehicles. 'Tracks' were dismissed because of the maintenance required for desert operations. As a result the Light Armoured Vehicle (LAV) was launched to provide the US Marines with greater mobility. The LAV 25 entered service with the Marines in 1983 and were first used in combat in 1989 in the Panama Invasion.

With the move of the Army to the North, there was renewed interest in wheeled armoured vehicles. I decided that it would be worthwhile looking at available vehicles of this type worldwide and made this a key area of interest in a world trip in July.

* * *

I arrived in San Francisco on the 16th July. The flight itself had been uneventful but I got lost in traffic trying to locate my hotel.

The next day, Sunday, was a rest day so I took a drive through the Nappa Valley and visited a Carmelite Monastery. It was hardly a pleasant day as the temperature reached 103 degrees, the hottest day ever recorded to that time.

My first working day was a visit to the FMC Corporation, San Jose, manufacturers of the M113 Armoured personnel Carrier and the Bradley Fighting Vehicle. I was intrigued by the history of the company. FMC (Food Manufacturing Corporation) started as a chemical manufacturing company making insecticides. At the outbreak of WW11, the company received a contract to develop and build an amphibious tracked landing vehicle and continued to diversify after the war. After a very interesting visit, I moved to Ontario, California, as a prelude to visiting the US Marine Base at 29 Palms.

At 29 Palms I met the LAV25 Programme Director who gave me an excellent briefing about the vehicle and its development. More importantly, he arranged for me to drive the vehicle in the Marines Test facility.

I was escorted my a young Marine Captain who gave me some instruction in the techniques of driving and then proceeded to put me in the driving seat with the injunction "Go for it, General!". Go for it I did and managed to get six of the eight wheels off the ground at one stage. This was a disappointment as eight wheels off the ground would have given me a golden wings badge. Driving the LAV was quite a challenge as I was thrown about quite a bit and resorted to keeping one arm in front of me to prevent my teeth being knocked out.

I completed the test run in great spirits and was probably a little enthusiastic driving back to the base as I was stopped by a big burly Marine Military Policeman. My escort tried to deal with the problem with much gesticulating and references to a visiting Australian General. The MP was not impressed and I'm afraid the Captain got a ticket.

Next stop was the US Engineers Waterways Experimental Station in Vicksburg. I wanted to visit there as they had developed an advanced mobility model and were experimenting with driverless vehicles. The mobility model was particularly interesting as we were working on a similar system in EDE.

I had to sing for my supper as I had been asked to address the 4/12 Reserve Officers Association. I was given a good hearing about the work at EDE but most of their questions related to my St John decoration, Serving Brother of St John!

The next day, I flew to Washington and had an opportunity to meet with the Head of Defence Staff and was able to catch up with Lieutenant Colonel Danny O'Neill and Luke Ward from RACAL.

Back to looking at armoured vehicles, I visited Detroit to view the manufacturing line for the LAV, General Motors Canada and the

Canadian Land Engineering Test Establishment. Of particular interest was the experimental six wheel variety of the LAV.

Next stop was London for a short stay.

My first visit was to the Military vehicles and Engineering Establishment in Chertsey, Surrey. Much to my surprise, the Army Liaison officer was Colonel Mike Rose, a fellow student at Staff College, Queenscliff, in 1971. While there I had a good look at their test track and saw the Chieftain tank going through its paces.

Saturday was a rest day and I had lunch with Brigadier John Paley at the Cavalry and Guards Club. John had been an exchange officer at RMC when I was a cadet. We were able to go through some details of his planned visit to Australia in December to attend the thirtieth reunion of our graduation from the College.

On Sunday I was able to catch up with my classmate, Bob O'Neill, Titular Professor at All Souls College.

On Monday I had a short visit with Plessey and then Computer Analysis and Programmers Group. I have no memory of the reason for visiting CAP and my diary was of absolutely no help. Following this it was off to Vienna as a prelude to visiting Steyr.

I had visited the Steyr Rife Company previously but this time I was interested in the Pandur as a wheeled APC. These vehicles had impressive specifications and came in 6x6 and 8x8 versions. The Pandur 1 was purchased by the US Army for its Special Operations Command.

Next stop was to Berne, in company with Brigadier Ian Hearn, HADS London, to visit the MOWAG company that produced the Piranha family of armoured fighting vehicles. This family of vehicles have 4x4, 6x6, 8x8 and 10x10 versions and, like the Pandur, have an impressive set of features.

I thought that both the Pandur and Pirhana vehicles were worthy of further consideration and reported as such on my return to Australia.

This trip was exhausting due to the amount of travelling in the three weeks that I had been away. I thought the trip was worthwhile but my recommendations regarding AFVs did not get much support as the LAV 25 was much favoured by the Minister.

* * *

In September, I was asked to host a visit by a Royal Thai Army General and his party to EDE. I must say that my staff and I expended a lot of effort in assembling an interesting programme of briefings and visits to the various engineering sections but a lot of this went for naught as we had great difficulty moving the general from the photography suite that sported a disappearing corner (curved plastering that provided a large uncluttered background for photographs) that he thought would be extremely useful for interrogation purposes!

* * *

As the year was drawing to a close, I had a call from my RMC classmate Bill Coburn (senior officer postings in Defence) to see if I would be interested in becoming a Visiting Fellow at the Australian Defence Force Academy(ADFA). As I now had only three and a half years to go before retirement and promotion was unlikely, I thought the proposal rather intriguing.

Visiting Fellows at ADFA had been introduced soon after its establishment as a means of encouraging students in their academic studies – we were to act as 'role models' and demonstrate that a good degree was most useful to a military career.

Anyway, off I went to Canberra to be interviewed by the Rector, Professor Harry Heseltine, Head of Electrical Engineering, Professor John Richards, and two other academics. The interview was a requirement of the University of New South Wales to ensure that the interviewee had appropriate academic qualifications and was a suitable person to be let loose on students. All went well and a few days later I was offered an appointment to commence with the 1989 academic year. Later John Richards told me that they were pleased they had passed the interview!

* * *

Back at EDE I started tidying up loose ends of projects that I wished to see finalised before I handed over to my replacement.

One item of note that occurred in November was attendance at one of the evening performances of the Army Tattoo. I had been asked to take the salute at the end of the evening so was resplendent in Mess Dress with miniature medals. At the end of the performance there was a small party for the performers that Shirley and I attended. During the evening a young member of a visiting Canadian Armed Forces Band approached me with a look of great admiration and said, looking at my St John Cross:

"Sir, I've never seen Victoria Cross before". I replied "You still haven't" and explained that the VC was Bronze on a purple ribbon while mine was Silver on a black ribbon. The poor lad was quite disappointed!

So now started the business of preparing hand-over briefing notes, preparation of inventories for the removal, farewell calls on the various headquarters and the odd party with the staff at EDE.

I had very much enjoyed my second posting at EDE and was pleased to see the end of some of my pet projects. However, it seemed time to go and I was looking forward to a new challenge.

30TH REUNION OF 1958 CLASS

I travelled back to Canberra in December for the 30th Anniversary of graduation from RMC.

It was a grand affair and we had pride of place on the terrace overlooking the parade ground along with the remnants of the 48 Class and the much larger contingent from the class of 68.

John Paley came out from the UK to join us and had a thoroughly enjoyable time. Still as eccentric as ever I thing he was flattered by the amount of attention he received. Some NZ members of the class attended but, I'm afraid, I only remember Keith Mitchell.

We followed the usual format: welcome cocktails; a memorial church service on the Sunday celebrated by the RC Chaplain of our time, Monsignor John Hoare; dinner on graduation eve; and attendance at the parade on Tuesday.

I never relinquished my appointment as President of the Quarter Bar Committee so I presided over the dinner with a large collection of suitable jokes. Our BSM, John Moody, gave an excellent speech as did Mike Jeffery, the only two-star in the class, and Bob O'Neill, our distinguished Rhodes Scholar, academic and historian. As usual, our class orderly, John Bullen, did a marvellous job in preparing brief, etc.

It was a most enjoyable gathering and I still marvel at how we all are able to take up conversations that had started years ago. Our class has always been a very tight group, no doubt in part because of our shared experience of the Lake George tragedy.

ADFA

"Six munce ago I cudn't even
Spel Akademick and now I are one."

I started at the Academy on the 6th February 1989 in the Department of Electrical Engineering. The Head of Department, Professor John Richards, a fellow alumnus from UNSW, was very helpful in my transition.

At ADFA, I started using an electronic diary using software that is no longer supported. I have very few paper records of my years there, so it is possible that events that I mention may be out of time sequence, for which I apologise.

Life at ADFA was quite a change in more ways than one. My large office, complete with a secretary, was replaced by a room not much bigger than a broom closet. Instead of a lieutenant colonel staff officer and other staff, I had access to the Departmental secretary for dictation and some typing. No longer did I have my Ford Fairlane Staff Car with a one-star plate, pennant and driver and not even a reserved parking space. I also found after the first week that I would take my place on the tea-lady roster and be responsible for keeping the milk, tea, coffee and biscuits available in the staff room. When John Richards heard of this, he said that this was not appropriate but I insisted that I take my turn with the other academic staff. My dedication to this task was acknowledged with a framed university tea towel when I left the Department some years later.

The staff of the Department met for morning tea and lunch in a room that doubled as a space for presentations, discussions, faculty meetings and Bridge. I had never played the game, having been turned off it early in my career by a group in the Mess at Watsonia. I was determined to 'give it a go' and undertook instruction from John Arnold and other members of staff. I was never very good at the game, but found it a great break from work in the middle of the day.

Another innovation for me was being re-introduced to golf. I joined the RMC Golf Club and joined with faculty members for nine holes after work on Fridays. I had played a little at Staff College but never pursued the game after that. This was different as I became a victim of golf obsession that has remained with me to this day.

My introduction to lecturing was with the third years. I was assigned to teach modulation theory as a gentle introduction to my new endeavour. I started out confidently with a good text book and a detailed curriculum. I soon learned that the task would have some challenges, as I found that I had to delve back to some basic trigonometry to become confident with the mathematics of modulation.

It was soon time for me to face the students and I guess I was as nervous as they were. I wore uniform and was accorded the due deference for my rank with the class standing when I entered the room and being addressed as Sir – quite a different situation to that of the civilian lecturers.

I clearly remember my first lecture when I introduced myself as a graduate of RMC Class of 58. Much eye rolling and subdued gasps of breath that someone so aged could still stand upright!

I must say that I soon warmed to the student body. In the main, they were an attractive body of keen, fit and enthusiastic students as one would expect given the recruiting process. In fact, I could see many of my classmates returned to their youth.

I was greatly amused with the freshmen cadets in the first days of joining. They had uniforms but had not received any instruction on protocol, drill, saluting, etc and so were not quite sure how to react to an ageing brigadier complete with red banded cap and red tabs. They obviously knew they should do something, but had no idea how to proceed. My usual reaction was to salute and tell them to "Carry On".

* * *

Given that the pace of life was slower in the academic world, Shirley and I decided we would travel abroad as often as possible. I had also made a promise that our travel would be First Class, having seen a large poster in the Departure Lounge of Frankfurt Airport proclaiming:

"Fly First Class or your children will."

We decided to concentrate on the British Isles and Eire for this first substantial trip to include visits to John Paley, the Patrician Brothers in Eire and to Shirley's friends, John and Dorothy Wallace in Mauchline, Scotland.

We flew out of Sydney on the 25th of June, 1989, to Bahrain via Singapore. We had decided to spend a few days in Bahrain, as neither of us had any experience of the Middle East. We arrived in the early hours of the morning of the 26th and after a short rest, started our day with breakfast in our hotel. Our introduction to Bahrain was interesting to say the least. The hotel clientele included beer drinking cigar smoking Arabs in flowing white thawbs and keffiyehs or wearing well cut business attire, ladies in full length burqa with niqab through full length abaya with hijab to skimpy high fashion dresses. There were lots of liveried staff.

We spent the morning visiting the Heritage Centre and buying material, perfume and some jewellery in the Suq and then retired for the rest of the day.

Next morning I hired a cab for a tour of the Portuguese Fort, Dhow makers, basket and cloth weavers, the National Museum and the Barbar Temple. All very strange to us, but interesting.

Prior to leaving Australia, a friend of Shirley's suggested we contact Don Hepburn who worked in the oil industry. I was a little diffident but eventually called the number we had been given. I said that I understood Don Hepburn works for you. There was a pause and then the reply:

"No Sir. We all work for Don Hepburn. He is the General Manager."

He was pleased to hear of our mutual friend and invited us to dinner in a local restaurant. We were picked up in a chauffeur-driven limousine and deposited in a very smart Italian establishment that was very popular with the expatriate community and the foreign intelligence fraternity. Don pointed out the local operatives of the CIA, MI6, KGB et al. Apparently this suited everyone as it allowed for mutual observation without too much effort!

After being driven around the Island by Don's chauffeur, I attended a Rotary lunch. I received the usual welcome given to a visitor and was allocated a local host. I remarked to him on the absence of female members. He replied that it was culturally impossible to include women irrespective of decrees from Rotary International.

As is usual with rotary meetings, the meal was followed by a Sergeants Fine Session. As a visitor I was immune as all fines were in multiples of 10 Drachmas compared to my Club's limit of 50c.

Our flight was delayed so we spent a further night in our hotel at Qantas expense. The extra time was spent in the Suq buying gold and perfume. As an aside, neither Shirley nor I felt in the slightest concern for safety as we walked around the Bahrain streets at night.

* * *

We hired a car on arrival in London on 29 June and paid an insurance surcharge to take the car to Eire. We then drove via Windsor Castle to Salisbury and Stonehenge and arrived at John and Christina Paley's house late afternoon. I'm pleased to report that John was as eccentric as ever in addition to being a marvellous host.

Next day we drove West and spent two days in Wales before catching the ferry to Rosslare. We headed directly to the monastery at Tullow where the Patrician Brothers were founded in 1808. Our ferry had been delayed and we didn't arrive until just after midnight. I had called ahead and one of the Brothers had waited up to get us settled.

Shirley, being a staunch Anglican, had some reservations about going to Eire let alone staying in a monastery. She was quite amazed to find the co-operation that existed between the Christian faiths in Eire having read stories of sectarian violence in the past. As an example, daily Mass in Tullow was celebrated in the Church of Ireland church as the roof of the catholic church was under repair.

Shirley quickly warmed to the Brothers and thoroughly enjoyed their hospitality and humour. She often recounted the exchange she had with one of the Brothers who, when asked why he was not attending the retreat with the others, responded with:

"They are much too holy for me. I'll just sit quietly and smoke my pipe."

We toured around the southern counties of Eire for a few days visiting Galway (where we watched the sun go down over the bay), the Ring of Kerry, the Dingle Peninsular (lunch at Doyle's Fish Restaurant in Dingle), the Cliffs of Moher and the Rock of Cashel.

I was anxious to make a visit to the Cistercian Abbey of St Joseph in Roscrea as this was the Mother House of Tarrawarra Abbey in Victoria. I was unsure of the way and stopped one of the locals to ask for directions. Not far down to road was a 'T' junction and the local said go to the junction and Turn Right but he was holding out his left arm. I started to respond but Shirley whispered "don't be rude and do what he says". I thanked the man and proceeded to turn in the direction indicated and so started a long detour.

After a brief return visit to Tullow, we drove to Dublin to stay in the Stephen's Green Club. While in Dublin, we had the opportunity to view the Book of Kells in the Pro Cathedral.

Thus ended our visit to Eire. I had told Shirley how green the countryside would be but it was a bit of a disappointment compared with what we saw in Scotland.

* * *

We took the ferry to Holyhead on the morning of the 7th of July and proceeded to Caernarfon Castle and thence to Porthmadog for an overnight stay in a B&B.

Next morning we took a round trip on the Ffestiniog Railway. I booked First Class seats that turned out to be large lounge chairs in a separate carriage – they were certainly comfortable and we had good views of the slate minings.

After an overnight stay in the Lakes District we drove to John and Dorothy's house in Mauchline via Gretna Green. Mauchline is at the heart of Robbie Burns country and houses a number of shrines and museums of the bard. I must admit that the countryside was more impressive than Eire with the heather in bloom and green pastures.

John was working so I made a visit to the local pub where the only one I could understand was the barman – the rest were speaking, to me, unintelligible broad Scots.

Another day I took a trip to Kilmarnock for a haircut. I was intrigued to observe the barber and the patrons listening to the England–Australia Test Match. I innocently remarked that I didn't think the Scots were interested in cricket and was told "We are interested in any game where the English are being beaten!"

After leaving Mauchline, we travelled West to Glasgow took the ferry from Oban to Mull and then another ferry to Iona – a very special place of pilgrimage. Back on the mainland we made a clockwise trip around the North of Scotland with visits to the Pringle Factory, various distilleries, Pluscarden Abbey(then in the process of restoration that has since been completed), Braema, Inverness and Edinburgh.

From Edinburgh, we drove southwest and visited St Ninian's Cave near Galloway Bay before visiting Newcastle, and Durham. Proceeding South we visited York (magnificent cathedral) and then late in the day to Cambridge and finally, on 23 July, to the RAF Club in London.

While in London we were able to visit St John Gate, the headquarters of the Venerable Order of St John, took a boat ride to Greenwich and had dinner with old friends from Plessey.

* * *

We broke our homeward journey in Bangkok and called Lachie Thompson who was the Defence Attaché. No sooner had we made contact when he informed me that we would be playing jazz at the Swiss Embassy that evening. When Shirley said she had nothing appropriate to wear, Lachie called a local seamstress who came to our hotel, measured every conceivable feature of Shirley's body and promised a dress would be delivered by 6:30 pm that day. True to her word, she arrived at 6:25 with a perfectly fitting silk dress. I must say that I greatly enjoyed playing with Lachie again and it was a great ending to our trip.

* * *

Back to work on 1 August preparing lectures for the second semester.

Being now well and truly settled back in Canberra, my level of involvement in Rotary, St John Ambulance, the Joint Staff College Association and the Duntroon Society increased dramatically. I also kept in close contact with the Joint and Single Service Communications Directorates to ensure I was best able to keep my students up to date with current programmes and projects.

It was about this time that I commenced work to develop a series of lectures on unclassified cryptography. I became particularly interested in Public Key Cryptography in addition to the usual forms of system security as used by commercial as well as military systems. I regularly joined in cryptography seminars conducted by Dr Jenny Seberry who was the Head of Computer Science at ADFA before moving to Wollongong.

* * *

On the home front, I was now able to develop my interest in woodworking and proceeded to invest heavily in tools and books. I never did become very skilled in woodworking but I did get a lot of satisfaction from making, what my wife called, the most expensive saw dust in Canberra!

* * *

I finished my first year at ADFA attending the Graduation Parade and Conferring of Degrees. The parade was impressive, particularly as little time was available for drill and rehearsals. For the conferring ceremony I was part of the Academic Procession resplendent in White Ceremonial Uniform and wearing the gown and hood of a Master of Engineering Science. It was a splendid affair and I very much enjoyed meeting the parents of my students.

1990

My proposed courses were accepted and I now taught classes for the Third and Fourth Years and supervised some Four Year projects.

Mike Jeffery had arranged a travel budget for me so I was able to make various visits to keep up to date without calling on the limited travel funds of the Department. Mike and I kept in close contact as we were the only remaining serving members of the Class of 58.

I made a number of visits to EDE, DSD, the School of Signals and various communication facilities of the three Services and was always well received.

During the year I was invited to take part in the ADFA Tour of Defence Establishments around the country including HMAS *Stirling*, *Learmonth, Darwin* and *Townsville*. These visits were very beneficial to the students and reminded them of the ultimate goal of their studies.

In March my daughter, Kerry, married Lennon da Pozza in my old parish church in Curtin. My relationship with my ex-wife, Yvonne, had warmed slightly so the reception was reasonably pleasant.

* * *

Life was pleasant and busy with challenging work at the Academy, involvement with the Duntroon Society, Rotary, St John Ambulance and with relaxation with woodwork and golf.

My woodworking skills were improving and I had some minor success with frames for pictures and large mirrors plus a variety of boxes. I did spend a lot of time making jigs and fixtures and lots of sawdust and I managed to keep all my fingers. I must admit that I spent a considerable amount of money on tools, some of which were never used.

I was a member of the Board of my Rotary Club and I was also involved with Shirley in her Inner Wheel activities.

I had been appointed Deputy Commissioner of St John Ambulance Brigade in the ACT and was heavily involved with officer and NCO training as well as being a member of the ACT Committee.

* * *

I was brought violently back to earth in July when my good friend and classmate, Tony Hammett, died in an aviation accident. After leaving the Army, Tony established a single plane aviation business. He had been operating commercially for some time but his luck run out when he crashed soon after take-off with a full fuel load and five passengers. All were killed in the impact or the fire that followed.

In order to spare Tony's widow and sons, I was asked to travel to Brisbane the day after the accident to identify the remains. As it happened the task was beyond me and identification was only achieved through dental records. It was a traumatic experience for me and most distressing.

Tony was much admired by the Army in general and the Infantry

Corps in particular and he was given a Military Funeral. This was unusual as this honour was usually reserved for serving members and retired Generals. Members of the Class of 58 provided the Pall Bearers.

Tony had been the Commandant of the Infantry Centre at Singleton and his contribution to the civilian population of the region was recognised by Tony being made the first Freeman of the Shire.

* * *

In October, Mike Jeffery, who had been CSM in 1958, organised a church service for Alamein Company of RMC. I cannot remember the detail but Mike asked me to address the congregation. I had read scripture from the pulpit before but had never 'preached'. It was quite a daunting task and, luckily, never had to be repeated.

* * *

As examinations approached, I joking said to the Third Years that I could be bribed but only if the bribe was appropriate. I had also told them of my desire to own a Porsche 911. The day before the examination, the senior student asked if I would be in the next day. I assured him I would be in at the usual time.

At 08:00 on the appointed day, a delegation of Third Years appeared at my office door and invited me to go to the entrance of the Engineering Building. There before me was a bright red 911 that was mine for the day. I immediately drove home to Garran to show Shirley what I would like for Christmas. I will not repeat her response!

I found out later that the Class had hired the 911 the evening before and had spent the night driving the car to and from the Cotter rather than studying.

* * *

I got good use out of my white ceremonial dress in December by attending both the RMC and ADFA Graduation Parades and the Conferring Ceremony.

And so 1990 came to an end.

1991

At the start of the year I was asked by Mike Jeffery, who was Deputy Chief of the Army at the time, to take part in the Gyngell Review Of Army's information systems. This involved visiting the Functional Commands and some major units. The main aim was to assess the various information requirements of the commanders and their senior staffs. Unfortunately, we were generally fobbed off to the practitioners who, in most cases, seemed unable to think outside of their own sphere of influence. All in all, it was somewhat disappointing as I had hoped we could have discerned the major requirements that would allow for development of a coherent policy.

* * *

Back to ADFA and my lecturer duties.

During the semester, a colleague, Dr Lal Goddard, and I gave an external course on secure communications to one of the Defence contractors located in Adelaide. This was a welcome change from the usual routine and enjoyable.

* * *

In May I attended the centenary celebrations at my old school, Holy Cross College. This was a most enjoyable event as a number of my old teachers had come back from retirement in Ireland to attend.

* * *

I now had just over a year to serve before reaching retiring age. I was anxious to spend my last year directly with the Army, so Mike Jeffery, who was then Assistant Chief of the Army-Materiel, arranged for me to be posted to Army Office as Information Systems Adviser. The Posting Order soon arrived and I bid farewell to ADFA and was taken on Army Office strength on 1 July 1991.

Colonel Mike Collins was allocated to me as an assistant and I was given a small staff to establish an office.

We spent considerable time speaking with the major software, database, information systems and hardware suppliers including IBM, Microsoft, Prime, SAP, British Aerospace, etc. In many ways, information systems were still an emerging technology and there were many examples of large projects being abandoned due to time and cost over-runs.

Towards the end of the year Mike Jeffery suggested that I should do a review of software support systems as operated by our ABCA partners.

Mike Collins and I commenced planning and negotiating visits to the allied agencies. Approval was given for us to take a round-the-world trip to visit the most important sites. Our itinerary was brutal as Army policy at the time set a limit of three weeks for 1-star trips. Luckily for me, Army policy allowed for Shirley to accompany me. The criteria for accompaniment were many and varied and it took some little effort on Mike Jeffery's part to obtain approval.

Shirley, Mike Collins and I set off westward to London with a rest day in Singapore on the way. In Singapore we three were well looked after by the Defence Attaché, Kerry Mellor.

As an aside, Kerry had a most interesting career with his most precious possessions were letters telling him he would not be promoted to lieutenant colonel, colonel nor brigadier!

Our hotel in London was the Hilton Olympia. Our expensive room was tiny but provided some unexpected amusement with a man in a neighbouring flat going naked as soon as he entered.

While in the UK, Mike Collins and I visited the Director General

Communication-Information Systems in the MOD and the software support facilities at Larkhill and Blandford.

After our four days in the UK we flew to New York City, arriving on the 28th.

The next location was Fort Monmouth, where we had meetings with Communications Electronics Command and the Centre for Software Engineering.

Then to Washington, DC, where we visited TRW, our Embassy and, in the Pentagon, the Deputy Chief of Staff for Command, Control and Communications and related directorates.

On the 5th, we flew to Quebec to visit the Defence Research Establishment, Valcartier and then to Ottawa to visit National Defence Headquarters.

* * *

On the 8th we flew to Tucson, Texas. On arrival, I presented myself to the car hire firm, the name of the firm escapes me, that had the contract with Qantas. The arrangement with Qantas was new and staff were not well acquainted with the travel vouchers. As a result we had been delayed in every location in the USA with each agent politely wishing us to "have a Nice Day". I related this to the agent in Tucson who was most apologetic and asked how we could be recompensed for our dissatisfaction. I immediately asked that our car be upgraded to a Cadillac. This was immediately agreed and we drove off in great comfort.

We drove to Fort Huachuca (Motto "From Sabres to Satellites") leaving Shirley at Tombstone on the way. A quick trip back to collect her as the Base Commander was eager to entertain us all for lunch.

I had a most embarrassing experience with the Cadillac that greatly amused Shirley and Mike. We were expected to fill the tank before returning the car to the agents so I pulled in to a service station close by and then the fun started.

I fully expected that there would be a switch to open the petrol filler cover but none could be found. I tried pulling on the cover to open it without success. I next opened the boot thinking the switch might be there-again no luck. Next I tried the Handbook that described an emergency opening system but I couldn't make that work.

So, in desperation, I drove the agents and explained my dilemma. The young girl on the counter gave me a withering look and took me outside to the car and simply pressed the cover and it sprang open. Much laughter all round!

* * *

I wanted to see what the Marines were doing with software support so we drove North to the Twenty-Nine Palms base in Palm Springs and from there to Marine Corps Base Camp Pendleton. Here Shirley greatly amused the Marines by remarking that their "Can't see me suits" were very different to ours.

From San Francisco, we flew to Hawaii for a most interesting visit with Commander in Chief Pacific Command, Control and Communications.

We arrived back in Canberra on the 15th quite worn out with travelling almost every day. Mike Collins and I were then kept very busy writing the report of the visit and briefing quite a few interested parties.

The remainder of the year was taken up with assisting in the development of an Information Systems Master plan, assisting Gyngell with his presentation to the VCGS and meetings with Microsoft, IBM et al.

1992

Back to work in January for my last six months in the Regular Army. I had completed most of the tasks I had been given so had an opportunity to spend time on hobby-horses.

Early in the Month, I had a call from a large IT company who asked for an interview regarding a job with them. This was a surprise as I had not really given much thought to life after the Army and I had not been putting out feelers. Anyway, I agreed and I had a meeting with three executives. I cannot remember what the job was but it attracted a substantial salary.

The interview was pleasant and ended with an offer being made that I said I would consider. As the meeting came to an end, I mentioned that I would turn 55 in the coming July. This obviously caused some consternation but nothing was said at the time.

The next day I received a call from one of the executives during which the offer of employment was withdrawn on the grounds that: "I lacked sales experience". Obviously my age was a barrier although they were not prepared to say so.

In February, I attended a second resettlement seminar. I think this was an Army initiative but it could have been Defence. It was worthwhile and covered some financial planning matters but, more importantly, some good advice on dealing with the stress of leaving a long-term career. I had not thought about this aspect too much but as I got closer to retirement I began to ponder on what had I done with my life. What had I achieved ? Had I made a mark on my profession? Had I contributed to the career development of those who had been in my charge? Should I have followed a different path? There were many regrets and, I guess, I became somewhat depressed. But not for long as I will explain later.

As part of the resettlement programme, I was given the opportunity to attend the University of New England to take a Company Directors course. This was extremely interesting and informative and, although I was never on a commercial board, I was able to apply this new learning in my membership of the Board of St John Ambulance, the Duntroon Society, and the Canberra Services Club (in fact I resigned from that Board as a result of the course).

* * *

During the year St John Ambulance, ACT, was honoured by a visit from the Grand Prior of the Venerable Order of St John of Jerusalem, the Duke of Gloucester. There were a number of functions but the highlight for me was being presented with the medal of an Officer of the Order of St John. At the investiture, the District Surgeon, Dr Frank Keillor, was admitted to the Order as were a number of other members of the Operations Branch.

* * *

Without much warning, I became interested, bordering on obsession, with the amount of paper used by the Army. I was unable to find any definitive answers about actual expenditure but I knew it would be substantial. As part of my informal enquiries, I visited the Ordnance Depot at Bandianna, specifically the Stationery Store. The store itself was very large with boxes and boxes of forms, handbooks, certificates, manuals, etc. One boxful of certificates that really struck me was a multicoloured one with an outdated Army logo that was for presentation to a musician who had mastered a second instrument! A good example of economy of scale gone wrong.

* * *

I was given the opportunity to make a number of farewell visits to Signals units and the Commands. These were all very enjoyable but the highlight was a return to EDE. As I was escorted around the establishment, mention was made of the number of improvements to facilities and projects that had been completed during my two tenures there. This really did wonders for my morale and alleviated many of my doubts about the usefulness of my career.

* * *

The year sped on. I was dined-out at the School of Signals and was given a lunch by the Chief of Army at Russell and another by my classmate, Mike Jeffery, at ADFA.

On the day before retirement, I attended the Quartermaster Store at Russell to return what Army property still in my possession. I was surprised to find that I had not returned 'Covers, Staff Duties in the Field' for which I was docked 50c!

On the same day, I transferred my superannuation from the Defence Force Retirement Death Benefit to the Military Superannuation Benefit Scheme. The change was most beneficial as my DFRDB contributions were deemed to be MSBS contributions and these far exceeded the amount that the scheme allowed. There were very few in this situation and it seems that 'the powers that be' thought that it was not worth the effort to change the legislation to correct this.

* * *

And so the day finally came when my service ended and I was transferred to the Retired List. So ended thirty-seven years and five months in the Australian Regular Army. I had very much enjoyed my career as I had been afforded great opportunities for education and travel but, most importantly, the great honour of being in command of servicemen and servicewomen.

I particularly enjoyed my time as Director of Communications, an appointment that allowed me to influence the direction the Corps was taking and to advance the careers of some who had, in my opinion, been overlooked for promotion. I was also successful in gaining approval for the award of the Princes Anne Banner.

Commanding the Engineering Development establishment was another highlight. I had the good fortune to be appointed while Major General David Engel was the Chief of Materiel. He gave me a free hand in running the establishment and was always ready to provide support when needed.

The appointment to Director-General Joint Communications-

Electonics, the peak appointment in the communications-electronics field, was both rewarding and demanding. My main regret in this appointment was that I was unsuccessful in promoting the concepts of information systems. The adoption of suitable policies in this area came later but I hope that I had planted the seeds.

APPROVED SYSTEMS

I had not given a great deal of thought to employment after leaving the Service so, with no looking around, I accepted a position with Approved Systems with the title of Manager, Secure Systems. Approved Systems were an agent and supplier of Apple Computers. Their main customers were schools and graphic artists and my job was to break into the Defence market. My springboard was the fact that one of the US Command and Control Systems was largely Mac based. We in Approved Systems were keen to pursue this aim against such competitors as Microsoft and IBM but there was not a lot of support from Apple Australia.

While working with the company, I formed my own consulting company. The result was PJA Evans and Associates. This was a fortunate decision and came in handy later on.

I enjoyed working with the Approved Systems people but I did not enjoy the work very much so, when things got tough for the company towards the end of the year, I resigned. There was no rancour involved, and I did some contract work for them in later years.

BACK TO ADFA

I had not much liked my brief association with industry and I certainly did not want to be a 'gun runner' so I was delighted to be asked by Professor John Richards to return to ADFA on a part-time basis. As the position was part-time, he could offer me an appointment as senior lecturer. This offer I accepted with great pleasure.

I had a few subjects to teach, but my major tasks were presenting the technical aspects of the Navy Principal Warfare Officers Course and the Young Officers Course at the School of Signals, Watsonia.

My working week was three days, so I had plenty of time to indulge in my woodworking hobby. I thought I might work professionally, so I gained the business name of PJA Fine Furniture. I sold nothing but made a lot of sawdust!

I have always had an interest in Cryptography – I wrote on this in my MEngSc thesis. I took up this field again when I was a Visiting Fellow at ADFA, as mentioned previously. In pursuit of this interest, I attended the international convention, Eurocrypt, 1993, in Norway. I thought it would be worthwhile extending this trip by going to the USA for the AFCEA Convention in Washington, DC.

As I started planning our travel, I came across a British Airways promotion offering a transatlantic flight in Concord at no extra charge on a First Class ticket. With that part settled, we looked at including the Orient Express to take us from Europe to London. Shirley was very fond of Venice, so, following the conference in Bergen, we travelled by train via Munich to Venice. After a week in Venice, we joined the Orient Express for the overnight journey to Calais, then via Hovercraft to Dover, followed by a Pullman service to London.

In keeping with the atmosphere of the Orient Express, I included a dinner suit in my luggage to blend in with our fellow passengers. I was not the only male so dressed and many wore false moustaches for the Hercule Poirot affect!

With no form of modernisation, Simplon restored the train magnificently. There were no showers, and each carriage had an attendant at each end feeding a chip-heater to provide hot water. The meals were memorable, as expected, and all-in-all it was a most memorable experience.

The hovercraft trip from Calais to Dover was fast and comfortable, as was the journey to London in the beautifully restored Pullman carriages.

The Orient Express was expensive but worth every penny!

After a few days in London, we taxied to Heathrow for our trip to Washington. A special lounge for Concord passengers awaited us at Heathrow where we were treated to fine wines and Hors d'oeuvres while comfortably seated in leather armchairs.

Time for boarding and we filed into our aircraft. Quite small inside, but superbly fitted out with grey leather chairs and fittings. We had just got settled when the captain, a close relation of Biggles, welcomed us onboard but advised there was a problem with some instrumentation specifically related to the air intake for the engines. He said it was probably an instrument fault but if it was not, we might have to take the trip, horror of horrors, sub-sonic. Not wanting to take the risk, Biggles decided for us to de-plane while a replacement aircraft was being prepared.

There was a little excitement as we exited as two gay men had had an argument and one had sustained a minor cut on his cheek. The Purser announced that cabin crew would separate them after re-boarding. This drew a scream of anguish with the injured person exclaiming, "but he is my friend!"

We were soon on the replacement aircraft and on our way. I must say it was exciting watching the airspeed indicator at the front of the cabin showing our progress past the sound barrier. One interesting fact was that the small round windows at each row became hot to the touch as the flight progressed. The flight was uneventful, and we landed safely in Washington.

* * *

The AFCEA Convention was most enjoyable as I met several friends from Fort Monmouth days. Shirley had a great time at the AFCEA Convention thanks to the extensive partners programme.

The trip was extremely enjoyable as well as being most informative for me and my interest in cryptography and current trends in military communications. We made these trips the focus of of travel over the next four years.

* * *

Not long after returning to Canberra, Colonel (later Major General) Ian Gordon contacted me to see if I would accept his nomination to become the Colonel Commandant of the Eastern Region. I took this to be a great honour and accepted with alacrity.

I was unaware that the previous Colonel Commandant had recommended an old friend, Barry Tinkler, to replace him. I had not lobbied for the appointment as some members of the Corps in NSW claimed and this group were to cause me some difficulty many years later.

* * *

In July, the Rotary Club elected me as President, a position that Shirley's late husband had previously held. It was an unusual presidential year as I spent much of it away from Canberra, but I contacted the Club each Wednesday during the meeting by mobile phone.

Towards the end of the year, I was part of a team that went to New Guinea to do some volunteer building work at a convent school in Papitalli, Manus Island. Our group transited through Port Moresby and I was shocked to witness the security measures required to combat the 'Raskals'.

Our team included a builder, Carlo Binutti, who was a master of improvisation and, under his supervision, we built a septic system and completed the construction of a two-storied dormitory block.

There were two medical doctors in the group and they earned their keep by checking the health of the students and the nuns who ran the place. Since I was an electrical engineer, they assigned me to supervise the wiring of the dormitory, and I have a photograph of myself hanging off a ladder while connecting the mains to the building. I was more than relieved when I applied power with no short-circuits or other dramas, considering my only qualification in the electrician trade was a knowledge of Ohm's Law! More to my line was some work on their very primitive computer system and switchboard.

The nuns provided our meals and, to try to return their hospitality, Carlo undertook to produce a 'proper' pasta meal. I shall never forget the look of horror when the native nuns saw the long strands of spaghetti as they only ever seen what came out of a can.

* * *

Back at ADFA, it was business as usual with the usual flurry with end-of-year exams and preparations for graduation.

As was usual, my RMC Class had an end-of-year reunion that I could not attend, as explained in the next section.

A NEW BEGINNING

I had struggled with alcohol abuse for many years with little success in overcoming it.

In my late twenties, I became concerned and sought advice from the Regimental Medical Officer. His first question was, "How is your career going?" I answered it was going well, and I was receiving excellent Confidential Reports. His response was, "So, what's the problem?"

Over the years, my concerns elicited similar reactions, but medical staff did not refer me to an outside service until late in 1989 (I am uncertain about this date and it could have been plus or minus a few years).

I followed the psychologist's recommendation of a 'controlled drinking' approach. During our consultations, I learnt the skills of self-hypnosis and received extensive advice on alternating water with drinks, etc. I was further advised that if this did not work, total abstinence would be the next step. I'm afraid I did not heed this.

The 'controlled' approach was moderately successful for a time, but I continued to be troubled and entered the state when I knew that alcohol was affecting my physical and mental health, but I could not imagine giving up completely. Then followed the depths of despair, but I continued the 'controlled drinking' approach.

As time passed, I reached the stage of rarely getting drunk but never being entirely sober. I was still functioning, but I withdrew from social events to save potential embarrassment.

Matters did not improve and one Saturday morning in November 1993 Shirley, in desperation, called my GP an ex-RAN Surgeon. After a brief discussion, he gave me some Valium and organised a visit to a Drug and Alcohol Specialist the following Monday. This consultation was also short, and he proposed I should seek admission to the Detoxification Ward of Woden Valley Hospital. I agreed and within a few hours, approval had been obtained and I was admitted as a patient.

I had absolutely no idea of what to expect but entered with a firm desire to do anything to fix my problem.

In the week that followed, I actively took part in presentations on alcoholism, engaged in group discussions, extensively read literature on the subject, and attended meetings of Alcoholics Anonymous – in fact, the Ward required all patients to attend AA meetings every day.

It was a roller-coaster week, and it did not take me long to accept that I was an alcoholic. I learnt about the disease and, importantly, accepted that I was not an evil person trying to become a saint, but an ill person trying to get well. I also accepted that my task was to stay away from the first drink one day at a time.

I took the AA 12 Step programme on board and started doing the steps and reading 'The Big Book' written by the founders.

At the end of the week, the staff considered I was a candidate for discharge as there was an urgent need for my bed.

Shirley came to collect me. I was an emotional wreck, full of remorse and uncertain of my reception. As I settled in the car, crying and apologetic, Shirley said, "Let us wipe the slate clean and start again". She was as good as her word and never raised my drinking behaviour again. As one can imagine, this was an enormous help in my recovery.

One of the 12 steps is to make amends to those we had harmed as far as we are able. I attempted this but there was a large group that I still worry about – those young officers who observed my drinking but also saw my career progress. To these, I offer my unrestrained apology.

As I write this, I have been sober for nearly thirty years thanks to the grace of God, the love and support of family and friends and the Fellowship of AA.

* * *

By the Monday after my discharge from hospital, I had reached sufficient emotional stability to contact Professor John Richards at ADFA and Colonel Ian Gordon to advise them of my hospitalisation and the fact that I was an alcoholic and to offer my resignation as lecturer and Colonel Commandant, respectively. Both expressed surprise at my news, rejected my offer of resignation and encouraged me to persevere. Their support and encouragement were of enormous importance to me.

Later in the year, Shirley and I went to stay with her relations in Teralba near Newcastle. I had also received an invitation from Mike Jeffery to attend his dining-out at the Infantry Centre, Singleton. I was still feeling very vulnerable and so drove to Singleton to see Mike and to explain why I would not be attending the dinner. He was very understanding, and like John and Ian, offered me support and encouragement.

And so ended 1993.

1994

My teaching load at the Academy was unchanged, so I have little to report on that aspect of 1994. I kept my interest in cryptography and attended Professor Seberry's weekly seminars. I had to be reasonably careful with these sessions to not disclose what I knew about current equipment and systems used by Australia and its allies.

In June 1993, my Rotary Club elected me as Chairman and I dedicated a significant amount of time to Rotary business in the first half of 1994.

As Colonel Commandant of RASigs in Eastern Command (NSW), I spent quite some time visiting units and coming to grips with my responsibilities. Apart from a squadron in Newcastle, the units of the Corps were in the Sydney Metropolitan area, so it was no great imposition to visit regularly. Later in the year, I became the Representative Colonel Commandant of the Corps and remained in that appointment until 2002. During this period, I also kept the Eastern Command appointment, and this led to some difficulties in later years.

My memories of 1994 are vague as I struggled with sobriety while being fully occupied with teaching, Rotary, St John Ambulance, the Duntroon Society and Colonel Commandant duties.

One bright spot was my purchase of a Porsche 924 Turbo from a Rotary friend. I nominated this as my company car and could maintain it with some taxation advantages. Many Porsche owners regarded this model as a ladies' car, as they viewed only the 911 as a true representative of the marque. I would not argue with this view but I can attest it was very fast.

Shirley was never happy in the Porsche as she did not enjoy looking up to the hubcaps of trucks. I persevered and kept the car for two years.

CERTA CITO 75

Early in the year, I realised we were only five years out from the 75th Anniversary of the separation from the Royal Australian Engineers and nothing much had been done. After consultation with Brigadier Mike Swan, who was Head of Corps, I formed a small group to plan the events for 2000.

The committee members I chose were Colonel Tom Davies, Majors Bill Wattam, Terry O'Brien and Walter Buchanan, and this group remained intact throughout the planning, implementation and finalisation of the project.

Very early in the planning stage, we came up with the project name, Certa Cito 75, and devised a logo for use on promotional material.

One of our first consideration was the possibility of having our Colonel-in-Chief, the Princess Royal, attending part of the celebrations. In this we were very lucky, as it transpired that Princess Anne was to attend a Save the Children Fund event in Australia in 2000. This fortunate circumstance allowed us to save the expense of bringing her and her entourage from the UK, so we only had to arrange some internal travel. A very co-operative and helpful Private Secretary enormously helped plan for this aspect.

The committee decided early on that we should include a symposium acknowledging the technical history and future direction of the Corps. To reduce costs of such an event, we formed a partnership with Software Engineering Australia and hired the Canberra Convention Centre.

I do not propose to go into detail of the work done by the Committee, but suffice to say it included negotiations with suppliers of logoed clothing and memorabilia and sponsors. This latter aspect was, disheartening as we experienced many disappointments with rejections from firms that had sold considerable equipment and services to the Corps.

As time passed, I contacted my counterparts in the Old Commonwealth Signal Corps and the US Signal Corps to invite them to attend the celebrations and to mark the occasion with an exchange of gifts.

I also contacted the Governor-General, Sir William Deane, in his capacity as Commander-in-Chief to request that he open the conference and to which he agreed.

The Committee, on legal advice, moved to form a Foundation duly incorporated in the ACT. I shall return to this topic closer to 2000.

EUROCRYPT
1995

In keeping with my interest in cryptography, Shirley and I went to France in May to attend Eurocrypt 95. We arrived in Paris via London and, after some difficulty with luggage, went to our accommodation at the French Army Officers' Club in Place St Augustine. This establishment operates similarly to our own Army holiday facilities, setting fees based on rank. The Club provided us with a comfortable and conveniently located place to stay while we visited the sights of Paris.

After a few days, we took the fast train to St Malo where the conference was to be held. The number of older ladies who were there with their 'nephews' in the Hotel des Thermes surprised us.

The conference was much the same as the one I attended in Norway. Some very interesting papers and some heated exchanges between competing academics.

As was my usual practice, I dragged Shirley to see the quite remarkable monastery of St Michelle. Poor girl saw more Catholic churches, monasteries and pilgrimage sites than was good for a committed Anglican!

From St Malo we went by ferry to Jersey to stay with some Rotary friends and from there to London for a few days before heading to Washington, DC, for the AFCEA Convention.

I have to admit that I enjoyed AFCEA more than Eurocrypt, as it enabled me to catch up with old friends from the military and industry.

1996

Nothing much stands out in my memory of 1996, although I know I was busy with my Rotary Club and the Duntroon Society. Apart from lecturing at ADFA, I still did the odd consulting job for Approved Systems and Microsoft with occasional contact with IBM.

I had one major task with Approved Systems that involved me as Project Manager for a contract with Attorney-General's Department. The acronym for the project was SCALE on the Web and it involved digitising statutes and case law to be readily accessible on an intranet. Approved Systems was involved as the project was based on Macintosh computers. The A-G's staff had a preference against Macs, and their cooperation could have been better. However, we managed to finish the project on time and on budget, and my remuneration provided me the opportunity to replace my Mercedes 180E with an E220 that served us well for the next sixteen years.

1997

1997 brought a real melding of business and pleasure with a trip to Berlin and Eastern Europe.

In early April, I accompanied Shirley to an International Inner Wheel Convention in Berlin. Shirley was not an official delegate but, as a District Chairman (sic), she was a senior member of the Australian party.

We started the visit with a Homestay with members of Inner Wheel in Koblenz. I cannot remember how we actually got to Koblenz but our Passports show we landed in London and from there it is a mystery.

Our hosts in Koblenz were both judges. This title intrigued me because our hosts were quite young. I have appended a few words on the German legal system.

"The judiciary of Germany is the system of courts that interpret and apply the law in Germany.

The German legal system is a civil law mostly based on a comprehensive compendium of statutes, as compared to the common law systems. In criminal and administrative law, Germany uses an inquisitorial system where the judges are actively involved in investigating the facts of the case, as compared to an adversarial system where the role of the judge is primarily that of an impartial referee between the prosecutor or plaintiff and the defendant or defence counsel.

In Germany, the independence of the judiciary is historically older than democracy. The organisation of courts is traditionally strong, and almost all federal and state actions are subject to judicial review.

Judges follow a distinct career path. At the end of their legal education at university, all law students must pass a state examination before they can continue on to an apprenticeship that provides them with broad training in the legal profession over two years. They then must pass a second state examination that qualifies them to practice law. At that point, the individual can choose either to be a lawyer or to enter the judiciary. Judicial candidates work at courts immediately. However, they are subjected to a probationary period of up to five years before being appointed as judges for life."

Among the husbands of Koblenz Inner Wheel, were a retired colonel and a major general. The colonel came with a duelling scar and a total deficit of humour. The general and his wife were charming and, I think, of the old aristocracy. There was certainly much evidence of old money. They hosted us to dinner where we ate the venison that our host had shot in Scotland a few days earlier.

From Koblenz we travelled to Berlin for the actual convention. I must say I enjoyed the role reversal in being 'the accompanying person'. There was an excellent programme that was obviously designed by an engineer as we took a trip in a vintage 'S' Bahn, visited a technology park and the boat lift at Niederfinow. The Niederfinow boat lift is the oldest working boat lift in Germany and allows for a 36m inclination between the Oder-Havel Canal, allowing access to the Black Sea.

* * *

After the Convention, we took the Post Convention Tour of Romantic Germany. The tour itself was to start at Heidelberg, so our first leg was to get there from Berlin via an Inter City Express train. The organisers displayed a high level of organization, as evidenced by:

    a. 0730. Meet at Station

    b. 0802. Train departs

    c. 1351. Arrive Heidelberg

    d. 1800. End of tour at Hotel

The tour was a standard CPO Hanser one but enhanced by Inner Wheel members, who met us at every stop to provide hospitality and fellowship.

First stop was Würzburg and thence to Tauberbischofsheim for a wine tasting in the castle cellar. Then to Creglingen and ending the day in Rothenburg, one of the oldest medieval cities in Germany.

Continuing South until finishing the tour in Fussenand, bidding farewell to our incredible Inner Wheel hosts.

* * *

To make the most of our trip, we booked a tour of Central Europe to start and end in Frankfurt. I mistakenly said we were touring Eastern Europe and someone promptly corrected me.

So, after a day in Frankfurt, our coach took us to Berlin via Basel and Brunswick. Two nights in Berlin and then Warsaw via Potsdam. The contrast between East and West was immediately obvious. Infrastructure needed a substantial influx of funds and there was ample evidence of poverty among the general population.

Two days in Warsaw allowed us to visit the Ghetto and to listen to the obligatory Chopin recital.

Our next stop en route to Kraków was Częstochowa to visit the Pauline Monastery of Jasna Góra and the famed Black Madonna icon. This was a highlight for me, but Shirley was much less impressed. In fact, she remarked, "So you have dragged me halfway around Europe to see this tiny painting." I confessed she was right.

From there to Auschwitz, a terrible monument to man's inhumanity to man, and then on to Kraków.

A highlight of Kraków was a visit to the salt mines. Over decades, miners had sculpted sacred figures and chapels in the salt.

From Kraków, we travelled through Slovakia to Hungary for two nights in Budapest. There was a mandatory visit to a horse farm for a breathtaking exhibition of horsemanship.

Two nights in Vienna and from there to Prague in the Czech Republic. From there, back to Frankfurt.

* * *

After all this pleasure, it was back to work at Eurocrypt '97 in Konstanz, Germany. There was an extremely busy programme for the next four days, but there was an opportunity to take a cruise on Lake Bodensee made famous by Count Zeppelin.

* * *

So back to the real world of lecturing at ADFA, a little consulting work and ongoing involvement with Rotary, St John Ambulance and CC75.

1998–1999

Nothing much changed in my teaching load at ADFA except that I became involved in some aspects of the Principal Warfare Officers' Course we ran for the RAN. It was really enjoyable to interact with students in an older age bracket.

In 1996, the members of St John Ambulance, ACT, elected me as Treasurer, and I was reasonably busy in this role, primarily devising methods to find out our actual costs in presenting First Aid courses and supporting volunteers.

St John recognized my service in mid-1998 when they promoted me to Commander Brother of the Venerable Order (CStJ) and I had the pleasure of being invested with that honour by the Governor of NSW.

I was also very much involved with the Duntroon Society along with my class-mate, John Bullen. We experienced some difficult periods and at one stage, we considered winding-up. Luckily we survived this challenge, and the Society has since gone from strength to strength.

Notwithstanding all the above, my major activity was with CC75. The Committee met regularly to discuss some of the finer points of the programme for the year, memorabilia and a parade to mark the visit by Princess Anne.

I spent less time than I should have on Eastern Region matters because my concentration was on CC75 matters. This became a problem when someone complained to the Chief of Army about the lack of an annual dinner. I recall that I had canvassed several units to take the lead on this without success. The Head of Corps, Brigadier Mike Swan, came to my aid and strongly supported me with the Chief.

I felt very disappointed with the way they made the complaint, as they didn't make any representations to me and didn't provide me with a copy of the complaint. However, I sought approval for separation of the Representative Colonel Commandant appointment from the Regional Colonels Commandant and this has stayed in place.

After a period of discernment, the Sovereign Military Hospitaller Order of St John of Jerusalem of Rhodes and of Malta admitted me as a Knight of Magistral Grace. The Order is a Lay Religious Order of the Catholic Church.

Planning for Certa Cito 75 was well advanced, so Shirley and I took a holiday in England to see old friends and new ones made through Rotary International Visiting Fellowship. During this trip, I made a visit to the School of Signals in Blandford Forum, where their Head of Corps played the role of our host.

Some old friends from Plessey hosted us while in London. They asked me if I would consider working for them but I declined because I found academe to be quite suitable for me.

* * *

The first major event of the year was the visit by Princess Anne to Simpson Barracks, Watsonia. Our Colonel-in-Chief officially opened the museum, which we had moved to the old satellite ground station building, before attending a parade of representatives of the Corps. I had the privilege of hosting the parade and the morning tea that followed in the Corps Mess.

* * *

So now it was time for the Seminar. It was a ticketed event and our committee became concerned at the slowness of uptake. Indeed, I had a recurring nightmare of escorting the Governor-General on stage to be met by an empty auditorium.

Brigadier Mike Swan came to the rescue and 'encouraged' attendance. He also found some funds that enabled some level of subsidisation of the tickets.

The seminar and other functions were a resounding success, as was the sale of memorabilia. So much so that we had an excess of income over expenditure (we could not say profit) of over $60,000.

The Foundation Board kept a tight hand on this money, as there were many requests for help with regimental and other projects that could have whittled away the funds without showing some substantial result.

* * *

To round off the year, I received an invitation to attend the 50th Birthday Function of the Princess Royal at Windsor Castle in November. I was more than happy to accept and fund the airfare personally. The Army supported my attendance and provided me with travelling allowances for accommodation and travel while in the UK.

The party was a tremendous affair with representatives of the over 300 charities/foundations/associations that were under the patronage of the Princess. Fortunately, I could present loyal best wishes from her Australian corps and enjoyed a brief conversation with her.

As the event progressed, I was standing with the UK Master of Signals and various members of Old Commonwealth Signal Corps, when the Queen and the Duke of Edinburgh appeared among us – what a great honour and pleasure that was.

* * *

So back to normal with teaching at ADFA, Rotary, St John Ambulance, RSL, Certa Cito Foundation, making sawdust and golf.

I had taken up golf again while at ADFA and joined the RMC Golf Club. I was enjoying my game and slowing improving my handicap when my old friend, Alby Morrison, suggested he sponsor me for membership at the Royal Canberra Golf Club.

I took up his suggestion and processed with an application with, I think, four sponsors who attested I was a suitable candidate. The Waiting List was two to three years, and I paid the substantial Entrance Fee.

While attending a vetting interview with a few other hopefuls, I asked about the policy that the Club would apply if an aspirant died while on the Waiting List. I was promptly advised that the Club would not pursue my estate for any outstanding money but there would be no refund. So, a good encouragement to keep fit and healthy until admission finally occurred.

* * *

In mid 2002, I handed over the reins of Representative Colonel Commandant to Brigadier Brian Edwards. At about the same time, the Department of Electrical Engineering at ADFA was experiencing funding problems, so I felt it was a good time to retire. Thus ends the journey of another half mile.

EPILOGUE

It is now June 2024 as I write this and I will try to mention the high and low lights of the last twenty-two years.

* * *

In 2002, the Committee of the ACT Branch of the RSL elected me as the President after serving as Deputy President for a few years. During my presidential year, I had the good fortune of being part of a delegation of Branch Presidents that visited Japan as the guests of the Japanese Government. It was interesting to see how successive governments had washed out the years of the Second World War. We also found it interesting that our visit only included naval establishments and meetings with members of ex-Navy associations.

Our first meeting was with the Midget Submarine Association. We met with several retired admirals who hosted us and showed us the Australian newsreel of the burial with military honours of the midget submariners killed in the attack on Sydney Harbour.

We visited the Hiroshima Peace Memorial that presented a very biased view of the bombing. Close by were the Japanese Naval Academy and Museum. The museum holds artefacts of Heihachiro Togo, Horatio Nelson, John Paul Jones and Isoroku Yamamoto plus a memorial to Kamikaze pilots.

I resigned from the presidency after only one year to allow me to give more time to Shirley.

* * *

I continued my work with St John Ambulance in various roles and the members elected me as President. In 2005, I had the great pleasure of being 'dubbed' by my classmate Michael Jeffery as a Knight of Grace of the Venerable Order of St John of Jerusalem.

I handed over the presidency to Colonel John Quantrill, who served the Order and the ACT Branch extremely well.

* * *

Shirley developed Vertigo that caused her great distress when flying, so we became ardent cruisers. After being introduced to cruising on a trip to Canada, we followed this with circumnavigating Australia and taking several trips to New Zealand, New Caledonia, Vanuatu and Fiji and finally a 42 day cruise covering Indonesia, Hong Kong, China, Korea, Japan and Guam. Ill health caught up with Shirley and that prevented further holidays at sea.

* * *

Shirley's health deteriorated, and I became her primary carer. Shirley's daughters had been urging us to 'downsize' for many years without success, and there was little prospect that she would agree.

Shirley was a member of the St David's Anglican Church at Red Hill and our routine was for me to drop her and then go off to Mass and return to collect her and have morning tea in the Community Room. On a fateful day, there was a notice in the Community Room that an apartment within the St David's Retirement Village was open for inspection. We joined a small group and after a short time Shirley said "I could live here". I responded with "Are you sure?" and when I got a "Yes" I cornered the salesperson and said, "We will take it."

Then we started selling Garran and moving to the Close where our apartment was directly under the one occupied by Brigadier Jimmy Shelton and his wife. Jimmy was the much loved adjutant at RMC when I was a cadet.

Doctors had diagnosed Shirley with Parkinson's Disease and were treating her with medication and exercise. Her health deteriorated rather rapidly and, after a longish hospitalisation, I was told that she would need full-time care. Luckily, a bed was available in the Baptist Care establishment known as Carey Gardens that was very near to the Close. It was, indeed, fortunate, as Shirley had been attending the Day Care Centre at Carey Gardens, so she already knew some of the staff. It was extremely helpful for me, as I could be with Shirley every day.

The disease progressed and Shirley died in the early hours of the 13th March 2020. Canon John Campbell, an old family friend, took the service for Shirley's funeral, which was held on the 19th March in her parish church. COVID had just struck but the full level of restrictions were not yet in place.

* * *

I continue to live in St David's Close and enjoy the friendship of the other residents. I continue to play golf but must now use a cart and play only nine holes. The exercise content is unsubstantial, but continued contact with old friends is like gold.

In early 2024, I re-connected through my classmate, Steve Hart, with an old friend Audrey, the widow of Neil Harris. We go back a long way and my son, Damian, was madly in love with Linda, Audrey's younger daughter, when they were pre-teens.

Early in the piece, I was collecting Audrey to go to lunch when one of her grandsons took a photograph and sent it to his mother with the caption "Grandma's boyfriend opening the car door."

We both thought that 'boyfriend' was beneath our dignity and now describe ourselves as companions. We found that 'just good friends' and 'friends without benefits' was also not appropriate, so please stick with *companions*.

www.ingramcontent.com/pod-product-compliance
Lightning Source LLC
Chambersburg PA
CBHW030919090426

42737CB00007B/246